Search User Interface Design

Synthesis Lectures on Information Concepts, Retrieval, and Services

Editor
Gary Marchionini, *University of North Carolina, Chapel Hill*

Synthesis Lectures on Information Concepts, Retrieval, and Services is edited by Gary Marchionini of the University of North Carolina. The series will publish 50- to 100-page publications on topics pertaining to information science and applications of technology to information discovery, production, distribution, and management. The scope will largely follow the purview of premier information and computer science conferences, such as ASIST, ACM SIGIR, ACM/IEEE JCDL, and ACM CIKM. Potential topics include, but not are limited to: data models, indexing theory and algorithms, classification, information architecture, information economics, privacy and identity, scholarly communication, bibliometrics and webometrics, personal information management, human information behavior, digital libraries, archives and preservation, cultural informatics, information retrieval evaluation, data fusion, relevance feedback, recommendation systems, question answering, natural language processing for retrieval, text summarization, multimedia retrieval, multilingual retrieval, and exploratory search.

Search User Interface Design
Max L. Wilson

Information Retrieval Evaluation
Donna Harman

Knowledge Management (KM) Processes in Organizations: Theoretical Foundations and Practice
Claire R. McInerney and Michael E. D. Koenig

Search-Based Applications: At the Confluence of Search and Database Technologies
Gregory Grefenstette and Laura Wilber

Information Concepts: From Books to Cyberspace Identities
Gary Marchionini

Search User Interface Design

Max L. Wilson

www.morganclaypool.com

ISBN: 9781608456895   paperback
ISBN: 9781608456901   ebook

DOI 10.2200/S00371ED1V01Y201111ICR020

A Publication in the Morgan & Claypool Publishers series
SYNTHESIS LECTURES ON INFORMATION CONCEPTS, RETRIEVAL, AND SERVICES

Lecture #20
Series Editor: Gary Marchionini, *University of North Carolina, Chapel Hill*
Series ISSN
Synthesis Lectures on Information Concepts, Retrieval, and Services
Print 1947-945X   Electronic 1947-9468

Search User Interface Design

Max L. Wilson
Swansea University

SYNTHESIS LECTURES ON INFORMATION CONCEPTS, RETRIEVAL, AND SERVICES #20

MORGAN & CLAYPOOL PUBLISHERS

ABSTRACT

Search User Interfaces (SUIs) represent the gateway between people who have a task to complete, and the repositories of information and data stored around the world. Not surprisingly, therefore, there are many communities who have a vested interest in the way SUIs are designed. There are people who study how humans search for information, and people who study how humans use computers. There are people who study good user interface design, and people who design aesthetically pleasing user interfaces. There are also people who curate and manage valuable information resources, and people who design effective algorithms to retrieve results from them. While it would be easy for one community to reject another for their limited ability to design a good SUI, the truth is that they all can, and they all have made valuable contributions. Fundamentally, therefore, we must accept that designing a *great* SUI means leveraging the knowledge and skills from all of these communities.

The aim of this book is to at least acknowledge, if not integrate, all of these perspectives to bring the reader into a multidisciplinary mindset for how we should think about SUI design. Further, this book aims to provide the reader with a framework for thinking about how different innovations each contribute to the overall design of a SUI. With this framework and a multidisciplinary perspective in hand, the book then continues by reviewing: early, successful, established, and experimental concepts for SUI design. The book then concludes by discussing how we can analyse and evaluate the on-going developments in SUI design, as this multidisciplinary area of research moves forwards. Finally, in reviewing these many SUIs and SUI features, the book finishes by extracting a series of 20 SUI design recommendations that are listed in the conclusions.

KEYWORDS

search, information seeking, user interfaces, user experience, interaction

Contents

Preface

The world of Search User Interface (SUI) design is moving very quickly. Within the period of writing this book, many things changed: Google removed the Wonder Wheel, Yahoo! removed SearchPad, and Google+ was released — attempting to crystallise the use of '+1' around the web. Google made a significant departure from their normal look and feel, by including black and grey bars across the top, already dating some of the figures in this book. Further, Google began placing the green URL above the snippets of text within their search results, breaking a decade-long trend, and returning up to 10 sub-results in two columns for the first result on many searches. Some real-time search websites shut down or changed their niche, and Google ended their fire-hose contract with Twitter. Twitter also fully integrated their search subdomain within their main user interface. There have been major changes in structure of the search industry too. Hewlett-Packard even began their exit from the hardware market and bought one of the largest Enterprise Search companies, Autonomy, only a few months after Microsoft bought and then nearly abandoned FAST. Another Enterprise Search company, Endeca, has also been bought by Oracle. While this book covers historical contexts, empirically proven design recommendations, and experimental systems, it also covers many of the current trends. The latter of these may have already seen major changes by the time this book is published, but hopefully the framework described will provide sufficient structure for understanding these and the many future changes to come.

Max L. Wilson
November 2011

Acknowledgments

It is hard to consider exactly who has, or in deed, who hasn't contributed to my ability to write this book. My wife has certainly both encouraged and endured the process, but many people have inspired me with their work, supervised me in my work, and guided me in discovering all that I have understood so far. Notably, I would like to thank m.c. schraefel who supervised my doctoral research in this area, and set the model for the kind of academic I am now. There are also many academics that have inspired me to enjoy human searching behaviour and information seeking theory, including Marcia Bates, Nick Belkin, and Gary Marchionini. Further, I would like to thank those in industry who have pushed me to ground my ideas in real-world use wherever possible. At a more practical level, however, I would like to thank my reviewers who did an astounding job at providing both conceptual guidance and editorial rigour. Finally, I would also like to thank Gary Marchionini for editing this series of books and Diane Cerra for helping me to maintain pace in getting it all finished.

Max L. Wilson
November 2011

Figure Credits

Figure 2.1
Saracevic, T., The stratified model of information retrieval interaction: extension and applications. In *Proceedings of the Annual Meeting of the American Society for Information Science (ASIS'97)*. Copyright © 1997 John Wiley & Sons, Inc. Used with permission.

Figure 3.1
Slonim, J., Maryanski, F.J. and Fisher, P.S., Mediator: An integrated approach to Information Retrieval. In *Proceedings of the 1st annual international ACM SIGIR conference on Information storage and retrieval*, Copyright © 1978 Association for Computing Machinery. Reprinted by permission. DOI: 10.1145/800096.803134

Figure 3.2
Belkin, N.J., Cool, C., Stein, A. and Thiel,U., Cases, scripts, and information-seeking strategies: on the design of interactive information retrieval systems. *Expert Systems with Applications*, Copyright © 1995 Elsevier Ltd. Used with permission. DOI: 10.1016/0957-4174(95)00011-W

Figure 3.3
Palay, A.J. and Fox, M.S., Browsing through databases. In *Proceedings of the 3rd annual ACM conference on Research and development in information retrieval*. Copyright © 1981 Association for Computing Machinery. Reprinted by permission.

Figure 3.5
McAlpine, G. and Ingwersen, P., Integrated information retrieval in a knowledge workersupport system. In *Proceedings of the 12th annual international ACM SIGIR conference onResearch and development in information retrieval*. Copyright © 1989 Association for Computing Machinery. Reprinted by permission. DOI: 10.1145/75334.75341

Figure 3.6
Anick,P.G., Brennan, J.D.,Flynn,R.A., Hanssen,D.R.,Alvey,B. and Robbins, J.M., A directmanipulation interface for boolean information retrieval via natural language query. In *Proceedings of the 13th annual international ACM SIGIR conference on Research and development in information retrieval*, Copyright © 1990 Association for Computer Machinery. Reprinted by permission. DOI: 10.1145/96749.98015

Figure 3.7
Teskey, F.N., Intelligent support for interface systems. In *Proceedings of the 11th annual internationalACM SIGIR conference on Research and development in information retrieval.* Copyright © 1988 Association for Computing Machinery. Reprinted by permission.

Figure 3.9
Pejtersen, A.M., A library system for information retrieval based on a cognitive task analysisand supported by an icon-based interface. In *Proceedings of the 12th annual international ACM SIGIR conference on Research and development in information retrieval.* Copyright © 1989 Association for Computing Machinery. Reprinted by permission. DOI: 10.1145/75334.75340

Figure 3.10
Koenemann, J. and Belkin, N.J., A case for interaction: a study of interactive information-retrieval behavior and effectiveness. In *CHI '96: Proceedings of the SIGCHI conferenceon Human factors in computing systems.* Copyright © 1996 Association for Computing Machinery. Reprinted by permission. DOI: 10.1145/238386.238487

Figure 4.4
Hearst, M.A. and Pedersen, J.O., Reexamining the cluster hypothesis: scatter/gather on retrieval results. In *Proceedings of the 19th annual international ACM SIGIR conference on Research and development in information retrieval.* Copyright © 1996 Association for Computing Machinery. Reprinted by permission. DOI: 10.1145/243199.243216

Figure 4.13
Ahlberg, C. and Shneiderman, B., Visual information seeking: tight coupling of dynamicquery filters with starfield displays. In *Proceedings of the SIGCHI conference on Human factors in computing systems: celebrating interdependence.* Copyright © 1994 Association for Computing Machinery. Reprinted by permission. DOI: 10.1145/191666.191775

Figure 4.21
Teevan, J., Cutrell, E., Fisher, D., Drucker, S.M., Ramos, G., André, P. and Hu, C., Visualsnippets: summarizing web pages for search and revisitation. In *Proceedings of the 27th international conference on Human factors in computing systems.* Copyright © 2009 Association for Computing Machinery. Reprinted by permission. DOI: 10.1145/1518701.1519008

And

Woodruff, A., Rosenholtz, R., Morrison, J.B., Faulring, A. and Pirolli, P., A comparison ofthe use of text summaries, plain thumbnails, and enhanced thumbnails for Web search tasks.*Journal of the American Society for Information Science and Technology.* Copyright © 2002, John Wiley & Sons, Inc. Used with permission. DOI: 10.1002/asi.10029

Figure 4.25
Hearst, M.A., TileBars: visualization of term distribution information in full text information access. In *Proceedings of the SIGCHI conference on Human factors in computing systems.* Copyright © 1995 Association for Computing Machinery. Reprinted by permission. DOI: 10.1145/223904.223912

Figure 4.27
Veerasamy, A. and Belkin, N.J., Evaluation of a tool for visualization of information retrieval results. In *Proceedings of the 19th annual international ACM SIGIR conference on Research and development in information retrieval.* Copyright © 1996 Association for Computing Machinery. Reprinted by permission. DOI: 10.1145/243199.243218

Figure 4.28
Spoerri, A., Infocrystal: a visual tool for information retrieval. In *Proceedings of the 4th conference on Visualization '93.* Copyright © 1993 IEEE. Used with permission. DOI: 10.1109/VISUAL.1993.398863

Figure 4.35
Robertson, G., Czerwinski, M., Larson, K., Robbins, D.C., Thiel, D. and Dantzich, M.v., Data mountain: using spatial memory for document management. In *Proceedings of the 11th annual ACM symposium on User interface software and technology.* Copyright © 1998 Association for Computing Machinery. Reprinted by permission. DOI: 10.1145/288392.288596

Figure 4.4
Hearst, M.A. and Pedersen, J.O., Reexamining the cluster hypothesis: scatter/gather on retrieval results. In *Proceedings of the 19th annual international ACM SIGIR conference on Research and development in information retrieval.* Copyright © 1996 Association for Computing Machinery. Reprinted by permission. DOI: 10.1145/243199.243216

Figure 5.2
Morris, M.R. and Horvitz, E., SearchTogether: an interface for collaborative web search. In *Proceedings of the 20th annual ACM symposium on User interface software and technology.* Copyright © 2007 Association for Computing Machinery. Used with permission. DOI: 10.1145/1294211.1294215

Figure 5.5
Bernstein, M.S., Suh, B., Hong, L., Chen, J., Kairam, S. and Chi, E.H., Eddi: interactive topic based browsing of social status streams. In *Proceedings of the 23nd annual ACM Symposium on User interface software and technology.* Copyright © 2010 Association for Computing Machinery. Reprinted by permission. DOI: 10.1145/1866029.1866077

Figure 5.8
Ringel, M., Cutrell, E., Dumais, S. and Horvitz, E., Milestones in time: The value of landmarks in retrieving information from personal stores, *Proceedings of INTERACT 2003*, Copyright © 2003 Springer. With kind permission of Springer Science+Business Media.

CHAPTER 1

Introduction

The idea that search systems are built to be used by *humans* implies that searchers actively *interact* with some *Search User Interface* (SUI) in order to find the information that they need. In fact, searchers cannot even submit a first search query without some form of SUI, whether its command line, selection based, spoken, or virtual reality. By providing access to information, on or offline, the provider must presume that somebody will search for it. Beyond simply submitting search terms and displaying results, however, we now consider browsing interfaces, exploratory search systems, sensemaking problems, and scenarios that involve learning and decision-making. These different modes of searching presume longer periods of more diverse interaction. Consequently, SUIs are evolving to become increasingly interactive, in order to support a broader range of flexible and dynamic search experiences.

Although, research and innovation continue to create new and experimental user interfaces, there are many established SUI design principles that we see on a day-to-day basis. Perhaps most familiar is the streamlined experience provided by Google, but many more exist in online retail, digital archives, within-website (vertical) search, news archives, company records, and beyond. Amazon, for example, provides a multitude of different search functions, such as a search box, category hierarchies, filters, sorting, and more. Together, these features make Amazon a flexible, interactive, and highly suitable SUI gateway between their searchers and their products.

As a gateway between searchers and information, there are many factors and disciplines that have an impact on the design of SUIs. A SUI is limited by the metadata that exists and the algorithms in place to process them. Further, a SUI is affected both by the design of how we interact with it, and the aesthetics of the visual design. A great new search algorithm can be rendered useless, or unused, by a frustrating or annoying SUI. A great SUI can be rendered useless by poorly managed data and slow algorithms. Consequently, creating a good SUI involves appreciating these multiple factors, rather than simply making improvements in one area. Each of these factors, however, can be driven by the design aims of a new SUI. The aim of this book, therefore, is to set the reader into a multidisciplinary mindset before providing a framework for thinking about the elements that make up different SUI designs, taking into account when and where they are typically used. Further, the book aims to consolidate a substantial review of SUI designs and SUI features, and extract from them a set of SUI design recommendations.

1.1   FROM INFORMATION RETRIEVAL TO EXPLORATORY SEARCH

Information Retrieval (IR) is an established field of research [9] that has, since information was first stored within computers, aimed to get information back out of computers. Traditionally, IR has been system-focused: given specific query term(s), find the most relevant results to return. To facilitate this system-focused view, an evaluation environment was created called Cranfield [37], and later embodied in the TREC conferences [71], that for a given dataset and certain queries, had known sets of relevant results. In this evaluation environment, different algorithms (e.g., [88, 161]) could be competitively analysed for how close they could get to returning this known set of relevant results.

As described further in Chapter 3, researchers began to realise that more specific queries led to more relevant results, and soon after that SUIs should encourage searchers to improve or evolve their searches. Refining or improving a query, beyond a single set, became known as Interactive Information Retrieval (IIR). IIR evaluation began to highlight the interplay between interaction and algorithm, where the change in the proportion of relevant results was evaluated between interactions.

A theoretical discussion of what makes a result relevant [166] highlighted that beyond the matching of terms to documents, relevant results related to broader contexts such as the task around the search and the intention of the searcher. If searching for 'Orange Phone,' for example, searchers may wish to have results about the mobile phone company, rather than pictures of phones with an orange colour. Further, if their intention is to buy a new phone, then transactional shopping results would more relevant to the searcher than information about the company or its coverage. If a searcher does not yet know which phone to buy, then reviews of different phones may be more relevant for the searcher, given the stage of their task. Consequently, the design of search systems became more aware of the Information Seeking process [53, 103, 121] in general.

Information Seeking (IS) is defined as the resolution of an information need [121], where searchers typically recognise the need, formulate a search, evaluate the results, and complete their task. It is within this broader view of Information Seeking that many modes of searching behaviour have been discussed. Bates described a browsing model called Berrypicking [11], which captured the behaviour of searchers who collect information about their information need from various sources. More recently, 'Exploratory Search' [202] has been used as a term to represent scenarios where searchers cannot simply perform a quick search, but depend upon multiple searches, analysis of results, comparison, synthesis, and evaluation [120]. Exploratory Search scenarios, therefore, might include: learning about medical problems, choosing between products, or planning holidays. Exploratory Search is related to activities such as Sensemaking [44, 45] and decision-making.

These more complicated and exploratory scenarios are what lead people to begin searching, and determine how people decide whether their search has been successful. This book describes a large number of SUI features and designs that support searchers with different kinds of intentions, depending on what or who the SUI is designed for. Google is primarily designed to help searchers find the webpages they are looking for, and so focuses on providing support for IIR style activities. There are elements or 'features' of the Google SUI, however, that help searchers to explore more or similar results, compare shopping products, and so on. Further, within online retail environments that often focus on helping searchers find new and exciting products, the emphasis is often more on the exploratory features.

1.2   OUR EVERYDAY EXPERIENCES OF SEARCH

The SUI that many people in the world[1] now see on a day-to-day basis is Google, and Figure 1.1 shows the SUI features it provided for searchers on its Search Engine Results Page (SERP) in early

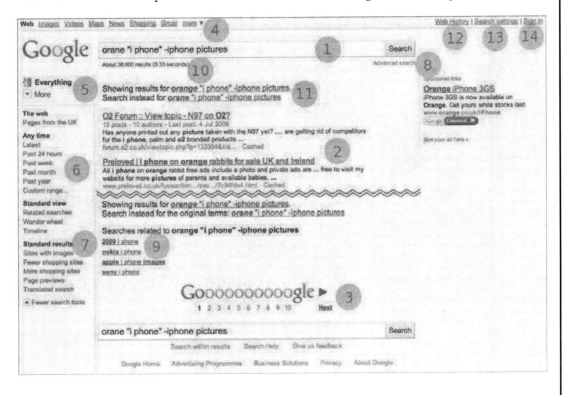

Figure 1.1: 14 notable features in the Google Search User Interface (SUI) in early 2011, described below.

2011. In one page we see 14 notable SUI features, although many more are hidden subtly within different types of searches and results. The most common of these features that searchers expect to see is the query search box (#1 in Figure 1.1). In Google the search box provides a maintained context of what has been searched for, so that the query can easily be edited or changed without having to return to the previous page. Searchers are free to enter whatever they like in the search box, including special operators that imply specific phrases (queries surrounded by speech marks) or make

---

[1]Although, China, which is quickly becoming the largest digital nation of the world, and several other eastern countries typically use Baidu.com; the choice and availability of search engines in China, are governed politically.

sure certain words are (with a $+$), or are not (with a $-$), included. The second most obvious feature is the display of results (#2), which is usually ordered by how relevant[2] they are to the keywords that were submitted. Results typically highlight how they relate to the search terms by highlighting, using bold font, the search terms within the snippets. Searchers are able to view additional results using the pagination control (#3).

We also see many SUI features in Figure 1.1 that help searchers to control or modify their search. Google lets searchers specialise to certain types of results, such as news or images, with a fixed set of filtering options across the top (#4). Further, a dynamic set of relevant options down the left (#5). Google also allows searchers to restrict the scope of their results (#6), or change how they are shown (#7). It is typical for search engines to provide an Advanced Search to help define searches more specifically (#8). Finally, Google, like most search engines, provides recommendations for related queries (#9).

Google also provides extra informative information, such as an indicator on the volume of results found (#10), and information about when searchers may have made an error (#11). Finally, Google also provides some personalisable features that are accessible when signed in (#14), such as settings (#13) and information about a searcher's prior searches (#12). All of these SUI features provide different types of support for searchers, which can be expressed in a framework.

1.3   A FRAMEWORK FOR THINKING ABOUT SEARCH FEATURES

In order to discuss SUI features and the kinds of interactive searching scenarios they support, this section provides an initial framework for thinking about SUI features, what they can be used for, how they are constructed, and how they should be evaluated. The framework below provides a language for discussing the design of SUIs and SUI features, which will be used throughout the book, and to structure Chapter 4.

Broadly, we can break the elements of a SUI, like those discussed in the Google example above, into 4 main groups:

- *Input* – features that allow the searcher to express what they are looking for.

- *Control* – features that help searchers to modify, refine, restrict, or expand their *Input*.

- *Informational* – features that provide results or information about results.

- *Personalisable* – features that relate specifically to searchers and their previous interactions.

To provide an example, these 4 groups are highlighted in zones over Google's SUI in Figure 1.2. This framework will be revisited throughout the book, as other SUIs provide different features in these groups. Often new SUIs, or SUI features, innovate in one of these areas, but it is important to

---

[2]The relevance of a result to a keyword search is in itself an extensive topic, and a loaded term. The idea of relevance is only briefly discussed within this book, but covered extensively in research and related literature.

note that many features can be used to provide support in all four groups. *Informational* features, for example, are often modified by *Personalisable* features, while other SUI features can provide *Input*, *Control*, and *Informational* support. The pagination in Google, for example (#3 in Figure 1.2), is both *informational* in subtly conveying the volume of results, but also allows people to *control* which page of results they see. Similarly, the auto-correction of the query (#11) is *informational* by conveying how a query has been modified, but provides *control* to searchers by allowing them to re-issue the original query unmodified.

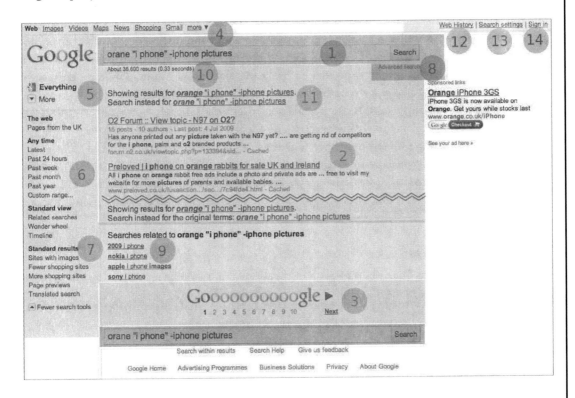

Figure 1.2: The Google SUI, from early 2011, zoned by the different types feature categories: Input as Red (including features #1 and #8), Control as Green (#4–#7, #9), Informational as Blue (#2, #3, #10, #11), and Personalisable as Yellow (#12–#14).

The description above has focused on Google as a case study of a familiar design for many. These features are representative, however, of other SUIs, like Bing, Yahoo! and China's main search engine: Baidu. It would be a relatively simple exercise to perform a comparison of the major search engines, to identify their SUI features and analyse how they contribute to one or more of the four elements of the framework: *Input, Control, Informational, and Personalisable.* Bing, for example, puts their dynamicaly list of specialised result types (#5) above the results but below the search box, as do

Yahoo!. Further, Bing provides a recent search history SUI feature on the left of the screen, rather than on a separate entirely on a separate page (#12). Baidu chooses to provide related searches at the very bottom of the page, like Google (#9), where Bing and Yahoo! place them on the left of the page. These and other feature designs are discussed throughout the book in terms of how they relate to the framework.

1.4   STRUCTURE OF BOOK

So far, this introduction has highlighted both the breadth of searching activities that this book intends to cover, as well as a framework for thinking about and discussing SUIs and their features. Before continuing to discuss SUIs using these contexts, this section first identifies a few key terms for reference. Chapter 2 then continues to bring the reader to a multidisciplinary perspective for thinking about what makes a strong SUI. Chapter 3 then provides a historical background to how SUI designs have changed and become established over time. Chapter 4 provides an overview of the many different SUI elements that we see both on the web and offline (although the evolution of the web has rendered these mostly the same). Chapter 5 begins to consider the future of SUI design with an introduction to more experimental designs that continue to question how we can best support searchers and their goals. Before drawing conclusions and collating 20 SUI design recommendations in Chapter 7, Chapter 6 describes an approach for planning how to evaluate the success of novel SUI designs.

1.4.1   KEY TERMS

- **SUI** – a Search User Interface.

- **SUI feature** – a single part of a Search User Interface (SUI) that provides a specific function, such as to query, refine queries, or display information.

- **IR** – Information Retrieval.

- **IIR** – Interactive Information Retrieval.

- **HCI** – Human Computer Interaction is the broader study of how users interact with software and technology.

- **Information Seeking** – the study of how humans achieve larger goals, which include (but are not limited to) IIR and other searching behaviours.

- **Exploratory Search** – a form of Information Seeking that is much broader than IIR, which involves higher-order activities such as decision making, learning, sensemaking, comparison, synthesis and evaluation.

- **Information Behaviour** – a term for overall behaviour with information, which includes seeking, but also creating, using, avoiding, and destroying information.

- **SERP** – a Search Engine Results Page.

- **UI** – a User Interface, or the bit that we see and interact with at the front end of any system.

- **UX** – User eXperience, a term for overall how a user will experience a system, including objective factors like usage time and mistakes, and subjective factors like aesthetics, usefulness and usability.

CHAPTER 2

Searcher-Computer Interaction

2.1 RELATED DISCIPLINES IN SUI DESIGN

As indicated in the Introduction, SUIs are heavily influenced by a number of factors: the data and the underlying systems, and the searchers and their situations. Saracevic [167] presented a layered model, shown in Figure 2.1, of the factors that affect SUIs. Saracevic's layered model essentially highlights that the SUI is a place where the searchers and the technology meet, and both sides have multiple factors that influence SUI designs. The user interface has to provide access to searchers who, based on their tasks, intent, and knowledge, have to produce a query. The interface is dependent, however, on what can be represented by data, and provided by the processes and hardware. More recently, Bates described a layered model of the components of a search system [10], but focused more on the many elements of the physical system rather than the searcher and their needs. Saracevic's model is perhaps ideal for representing the balance of complexities that exist on both sides.

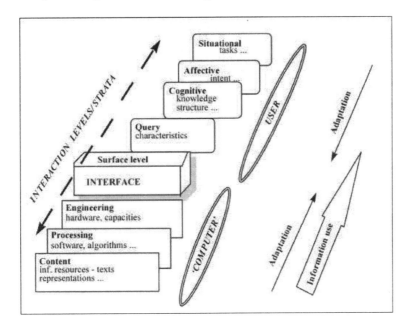

Figure 2.1: Saracevic's stratified, or layered, model of IIR, where the user interface is influenced by factors of the users and factors of the technology; taken from [167].

In practice, there are at least 6 disciplines, or factors, that all contribute to the design of a successful SUI design. These 6 factors, which exclude environmental constraints such as available hardware, finances, or the desires of management, are captured in Figure 2.2. From Library and Information Science, there are people that create, organise, and curate data sources and their meta-data. Whether a database, a corpus of documents, a curated collection, or community generated information, the available information has a limiting affect on what can be shown in a SUI, or how results can be manipulated. Searchers, for example, cannot filter a set of technical documents by the theme or subject, if thematic metadata has not been generated. Consequently, however, the structure of information sources may not only be driven by good Digital Library or Curation practices, but also how designers, or how actual searchers, wish to interact with the information. The design of good metadata and information repositories, therefore, is fundamentally tied.

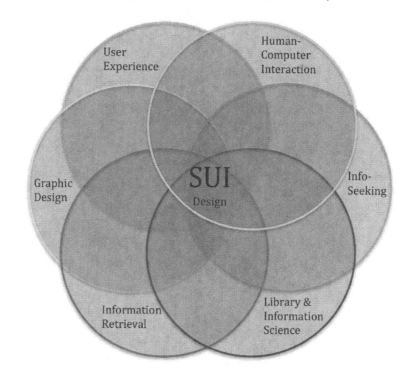

Figure 2.2: Search User Interface (SUI) design is affected by 6 related disciplines.

The available IR algorithms, and their efficiency, can also heavily affect a good SUI. A SUI cannot leverage clustered topics within result sets, if clustering algorithms had not been evaluated within the IR community [41]. In 2003, algorithmic response times, over the Internet, was noted as being one of the most significant factors affecting enjoyment of search engines [115]. Further, anecdotally, it is commonly understood that Google's major strengths lie in their significantly better algorithms for determining relevant results, as well as the speed at which they return them. Clearly, poor or slow algorithms can inhibit the success of a new SUI. Consequently, beyond an algorithm needing a SUI to be used, a SUI needs good algorithms to be used.

Beyond response times, simple graphical design elements, such as colour and layout, have been shown to have a significant impact on how people judge the trustworthiness of a website is [42]. Zheng showed that people judge how professional and trustworthy a website is within a fraction of a second, using purely aesthetic responses [221]. These kinds of findings indicate that simple aesthetic decisions made by graphic designers, or those creating the visual design, can have a significant impact on the success of a SUI. Further, many have shown (e.g., [15, 75, 137, 208]) that simple use of colour and other simple visual cues can have a significant impact on the success of a new SUI design or SUI feature.

The aesthetic design is one factor of a more broad view of User eXperience (UX). The UX community is primarily made up of practitioners who focus on designing, refining and completing a specific product or service. Hassenzahl and Tractinsky suggest that a user's experience is made up of 3 facets: temporal experience factors, emotional and affective factors, and holistic and aesthetic factors [74]. UX designers aim to make a UI simple to use, intuitive, and effective for a given set of users. Using methods such as wireframes, personas, and scenarios, UX practitioners create conceptual designs that meet the needs of users, and integrate effectively with their working needs. UX practitioners also evaluate working designs to remove elements that are confusing or frustrating in order to improve the experience that users have with the system. Work in 2002 [200] suggested that field trials and iterative evaluation methods, like the RITE method [125], were most popular, along with Task Analysis methods [46]. More recent work by Vermeeran et al. [199] catalogued 96 UX methods that are currently being used by UX practitioners. UX practitioners, however, are experienced in refining systems in general, to achieve a good user experience, but they may not have search-domain specific skills for nuanced SUI design.

Human-Computer Interaction (HCI) is a much broader discipline than UX that focuses on how people use computers to achieve goals and complete tasks. HCI has contributed many valuable approaches to evaluation, such as user studies, design methods, log analyses, eye tracking research, and many more that can be used to understand the human-side of search. It is typically the remit of HCI researchers to prove conceptually that certain UI styles are better than others for a given scenario of use, rather than to improve people's experience with specific products. Many of the advances in both general UIs and SUIs have developed from the HCI field, as seen in Chapter 4. Ideas such as Direct Manipulation (directly moving things within user interfaces, rather than using forms or options) [176] and Multi-touch interactions [112] were investigated carefully by the HCI community. Many SUI advances, such as clustering search results [151], and faceted navigation [208, 217], and use of tag clouds [91] have also been researched heavily within the HCI community.

Finally, the next community relating to the design of a SUI is one that thinks about how and why people search for information. By thinking more broadly about the themes of Information

Seeking (IS) or Information Behaviour, researchers have created many models and theories, such as: the types of search situations that people may be in [17], the stages that people go through when searching [53, 121], how people feel at different stages of searching [103], or the tactics people use when searching [12, 13, 177]. Fisher et al. catalogued 72 concise overviews of theories of Information Behaviour that together provide a more complete view over what is known about searching and related activities [59]. Although the IS community know a lot about how people search and why, they often use methods similar to those used in the HCI domain to evaluate systems with search-oriented tasks. Wilson et al. developed, however, an analytical evaluation method for SUIs that is specifically grounded in theories of IS behaviour [207, 211]. More about SUI evaluation is discussed in Chapter 6.

Each of these communities have had a significant and well-earned impact on the design and success of SUIs, but search systems are often designed or created with limited access to, or even knowledge of, each discipline. The SUIs discussed in this book come from each of these disciplines, and it is worth considering throughout the book: (a) which primary discipline has this SUI come from, (b) how is has it been impacted by the other disciplines, and (c) how can it impact the other disciplines. The search box, as we see later in this book, was motivated by algorithmic developments, has been styled by graphic design community, and modified by the UX community to affect the size of queries submitted. Further, the search box has become a persistent and consistent part of many SUIs, given its support for the majority of the Information Seeking process.

2.2   HOW WE SHOULD THINK ABOUT SUIS

It could be considered challenging, given the range of factors and disciplines described above, to consolidate exactly how best to approach the design of a new SUI. Ideally, however, systems should always begin with considering for what purpose they are being created, and for whom, so that every decision therein can aim to get it right for the user.

2.2.1   GETTING IT RIGHT – FOR THE USER

The aim of any UI, let alone a SUI, should be to understand and support the intended users as best as possible. Consequently, the design of SUIs is heavily influenced by the user-focused and the searcher-focused communities shown towards the top-right of Figure 2.2, and *facilitated* by advances in technical system-focused communities shown towards the bottom-left.

In designing systems to be right for the user, we can use HCI methods to learn about users, and frame them in the theories and models we have about IS behaviours. We can use HCI evaluation methods, however, to learn about how people use systems, or UX methods to refine and improve the experience with a specific product. It can be hard to learn about searchers of the whole web, however, as many different types of searchers perform many different types of tasks on search engines. In more specific domains, like online shopping, the tasks are more identifiable and the expected searchers may be narrowed down by the type of product. It may be easier to characterise, for example, the tasks for clients of a fashion retailer like ASOS[1], or the needs of clients of a computer components retailer like Dabs[2]. Essentially, however, if IS theory, HCI experiments, or UX evaluations suggest that the way searchers are supported is suboptimal, we are then able to consider how it should change and whether the underlying systems need altering in order to facilitate it.

2.2.2   THE EFFECTS OF THE TECHNOLOGY

Regardless of how we want to design SUIs to help users, SUIs will always be limited by the data available, and the algorithms that (a) analyse the data, (b) send the data to the SUI, and (c) respond to user interactions. These, however, are more under our control than the behaviour of human searchers. The need for faceted metadata (described below), for example, is typically scenario and opportunity driven. Google does not provide faceted search over its whole web search service, for example, but uses it purposefully for their product search[3], where the results are easily and logically classifiable by dimensions such as price, category, brand, size, and colour. In such a scenario, searchers may find a number of results that match their query perfectly well, but use these multiple dimensions of metadata to find the one(s) that best fit their needs, desires or preferences. When searching for websites, however, this second stage of manipulating result sets is less common.

Although ideally SUI designs should be user-focused and driven by recognised needs, in practice not all advances are SUI driven. Google's initial advances with the PageRank algorithm [25] were not driven by the desire to provide a new form of interaction, but by the optimisation or improvement of an algorithm to provide a faster and better service to users. On many occasions, as with clustering algorithms [41], the interactive advantages brought to searchers are proven sometime later [151]. Discovering the utility of algorithms for searchers, however, typically then drives the algorithmic development even further forward [194].

2.3   USER INTERFACE DESIGN PRINCIPLES

This book discusses many SUIs, SUI features, and SUI design recommendations provided by these six disciplines in more detail. This section, however, provides some initial insight into general UI design guidelines as produced by the HCI and UX communities. There are many UI design principles that can be adhered to, such as Shneiderman's 'Eight Golden Rules' [46], Mayhew's 'General Principles of User Interface Design' [30], or even Apple's Human Interface Guidelines[4] (should you approve of the aesthetics and usability of Apple software). Below, however, is a summary of 10 good general UI design principles, or design heuristics, provided by Nielsen et al. [137, 140]. Google is again used as a familiar case study, but these heuristics can manifest in many different ways in other search engines and SUIs.

[1]http://www.asos.com
[2]http://dabs.com
[3]http://google.com/products
[4]http://developer.apple.com/mac/library/documentation/userexperience/conceptual/
 applehiguidelines/

1. *Visibility* – Keeping the user informed about what is happening at any one time. Google, for example, maintains the current search terms in the query box (#1 in Figure 1.1). Further, if the SERP is not showing results directly linked to the submitted query, then it states clearly what corrected query it is using (#11).

2. *Language* – There are many technical things that computer software can do, but it is important to prioritise and describe what is happening in language that the user expects. Google, for example, stays away from the word 'query' and describes the query suggestions in a user-friendly way (#9) as 'searches related to [your search];' search professionals call these interactive query expansions or refinements.

3. *Control and Freedom* – It is important not to block users into a hole or fixed pathway, but UIs should instead provide users with the ability to easily recover from mistakes, or to change their plans. Many SUIs are good at noticing likely errors, such as spelling mistakes, but not forcing these corrections upon users.

4. *Consistency* – It is a good idea to follow conventions in design and consistently describe things in the same way. It would be a bad idea, for example, to call a query 'a query' in some places and a 'keyword search' in another place. Users may assume there must be a difference.

5. *Error Prevention* – Designers can help users by making it hard to do unproductive things. This is related to Control and Freedom except that UIs should try to help users avoid needing to undo their actions in the first place. This is perhaps why Google corrects a spelling mistake by default in many cases, so that the user only has to act when they did want to search for the unusual terms (#11).

6. *Support Recognition* – It is helpful for users not to have to remember what they have done or need to do. This is another reason why Google maintains the current query in the query box (#1). Similarly, suggestions are another way of allowing searchers to search using recognition.

7. *Flexibility and Efficiency* – Although many user interfaces are designed to be intuitive for first time users, it is also important to make sure expert users can do things more efficiently when they do not need the help. This is the reason many systems have shortcut combinations to print and save, for example, and why Google lets searchers navigate their search results with the up and down keys on a keyboard and use the return key to select a result (#1).

8. *Aesthetics and Minimalism* – Google has always maintained a very clean and minimalist design. Nielsen's principle recommends clever use of white space, to balance the amount of information being shown. Google's clean and clear design makes it easy to see exactly what to do next. For a brief period in 2010, Google choose to hide even the small amount of information that was visible on the front page, aside from the logo and search box, until the user moved the mouse.

9. *Clear error messages* – Many systems have error messages and these should be clear and informative to make sure the user knows what to do next. When users do reach a point of no results in Google, Google is quick to explain why and suggest alternatives. Errors that say 'there was an error' do not help the user to solve the problem.

10. *Help and documentation* – Providing clear help is important, even Google provides a page on 'Search Help,' as can be seen at the bottom of Figure 1.1.

Large online search systems, such as Google, Bing, eBay, and Amazon, have invested a lot of time and money to make sure they provide an efficient and intuitive SUI for the information they are providing. Undoubtedly, they apply the design principles above in many ways, asking if any SUI feature, as well as the whole design, can be improved. A number of examples have been provided above that relate to the Google SUI, but readers should think about these principles as examples are discussed throughout the book. Regardless of the specific search designs, tasks, and applications, these kinds of general UI design guidelines can be applied to refine future SUI designs.

2.4 SUMMARY

The section above has provided an overview for how to think about SUI design and SUI features, in the context of 6 different disciplines that, at the very least, affect the way search systems are designed. Although often bounded by the technology available, SUIs should be designed to support searchers as effectively as possible, rather than be focused on the underlying technology available. Short of describing all the techniques the HCI community has designed for trying to understand searchers, a set of 10 clear design principles have been described to help designers think about SUIs as we continue. The next chapter continues by describing the history behind some of the SUI design ideas that we work with today.

CHAPTER 3

Early Search User Interfaces

3.1   A BRIEF EARLY HISTORY OF SUIS

The roots of Search User Interfaces (SUIs), and indeed all Information Retrieval systems, can be found in Library and Information Science. In libraries, books are typically indexed by a subject-oriented classification scheme, and to find books we interact with the physical spaces, signposting, and librarians within them. In fact, one of the earliest multi-faceted metadata schemes was the Colon Classification for library books [156]. Yet the study of Information Retrieval was motivated by the development of computers in the 1950s, which could automatically perform one of the tasks that librarians may do: retrieve a document (or book). At the time, however, computers were controlled with punch cards, and later with text-only command lines interfaces. Systems were being driven by a clear model of ideal support (a librarian) but were so far limited by technology and method of interaction. Further, these early systems were developed before widespread access to the Internet, where searching was thus limited to a controlled and known set of documents that could be indexed like a library. Early forms of the Internet, such as ARPANET, were developed to connect specific institutions. Consequently, many early 'Online' systems, such as Orbit and DIALOG (both were evaluated for ease of use by Krichmar in 1981 [102]), were trying to provide librarian-style support over a fixed known library of digital records, but to remote machines that were physically distant.

3.1.1   CONVERSATION AND DIALOGUE

Given the UI limitations, and the influence of librarianship, some of the initial SUIs were modelled around conversations or 'dialogues' [143]. In analysing, for example, the roles, questions, and answers that took place in conversations between visitors and librarians [182, 212], early researchers developed question and answer style SUIs. In fact, these dialog style SUIs were often modelled upon the conversations between people with information needs, and the search experts (often called 'intermediaries') who were trained to use computers and early search systems [187]. dialogue-style system introduced in the 1970s [179]. These SUIs typically asked the searchers for any information they already had about what they wanted, and what they wanted to use the information for, so that when it came to performing the search (which could last a number of minutes or hours even) it was more likely to return useful results.

   This conversational style was analysed for some time, and was also influenced by those interested in Artificial Intelligence and Natural Language Processing, but the trend became less popular in search systems after technology (better algorithms and increased speed) improved and IR systems became increasingly interactive. It became less important to get exact information prior to searching,

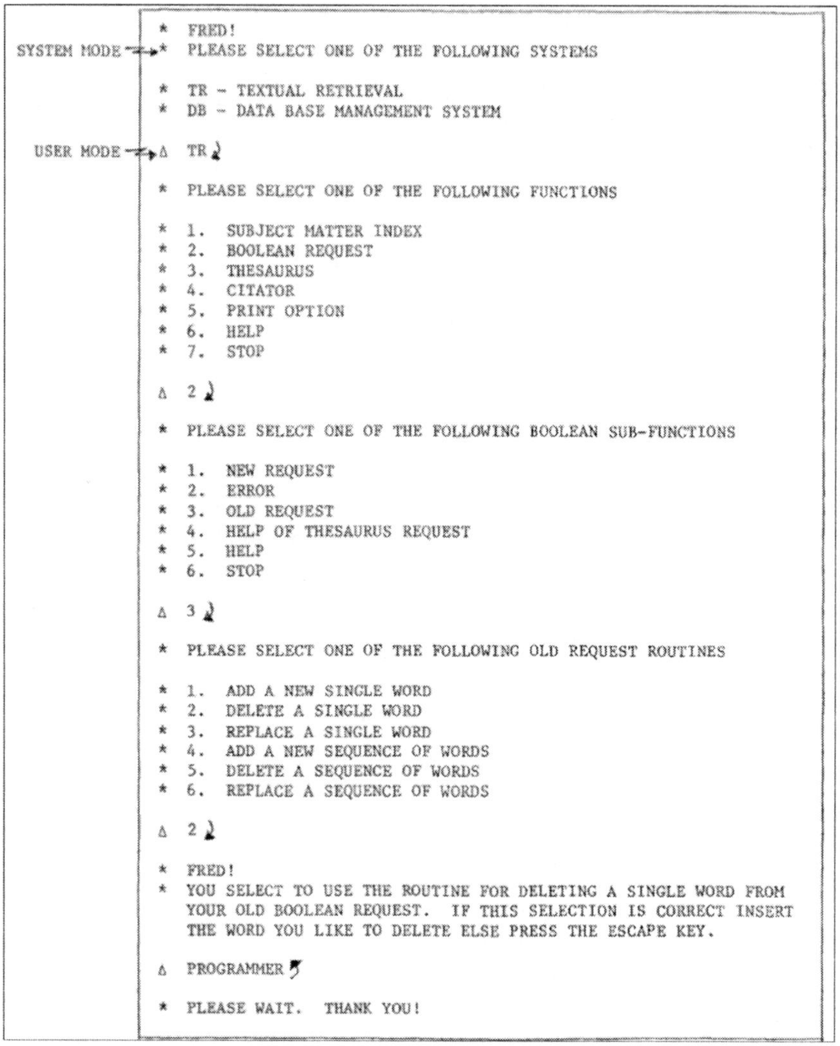

```
                    *   FRED!
SYSTEM MODE →→*   PLEASE SELECT ONE OF THE FOLLOWING SYSTEMS

                    *   TR - TEXTUAL RETRIEVAL
                    *   DB - DATA BASE MANAGEMENT SYSTEM

 USER MODE →→Δ   TR ⤸

                    *   PLEASE SELECT ONE OF THE FOLLOWING FUNCTIONS

                    *   1.  SUBJECT MATTER INDEX
                    *   2.  BOOLEAN REQUEST
                    *   3.  THESAURUS
                    *   4.  CITATOR
                    *   5.  PRINT OPTION
                    *   6.  HELP
                    *   7.  STOP

                 Δ   2 ⤸

                    *   PLEASE SELECT ONE OF THE FOLLOWING BOOLEAN SUB-FUNCTIONS

                    *   1.  NEW REQUEST
                    *   2.  ERROR
                    *   3.  OLD REQUEST
                    *   4.  HELP OF THESAURUS REQUEST
                    *   5.  HELP
                    *   6.  STOP

                 Δ   3 ⤸

                    *   PLEASE SELECT ONE OF THE FOLLOWING OLD REQUEST ROUTINES

                    *   1.  ADD A NEW SINGLE WORD
                    *   2.  DELETE A SINGLE WORD
                    *   3.  REPLACE A SINGLE WORD
                    *   4.  ADD A NEW SEQUENCE OF WORDS
                    *   5.  DELETE A SEQUENCE OF WORDS
                    *   6.  REPLACE A SEQUENCE OF WORDS

                 Δ   2 ⤸

                    *   FRED!
                    *   YOU SELECT TO USE THE ROUTINE FOR DELETING A SINGLE WORD FROM
                        YOUR OLD BOOLEAN REQUEST.  IF THIS SELECTION IS CORRECT INSERT
                        THE WORD YOU LIKE TO DELETE ELSE PRESS THE ESCAPE KEY.

                 Δ   PROGRAMMER ⤴

                    *   PLEASE WAIT.  THANK YOU!
```

Figure 3.1: A series of dialogue-style questions aimed to help the searcher describe what they are searching for; taken from [179].

as the wait times decreased. The ideas behind conversational-style interaction, however, still have an influence on IIR, although the 'conversation' now happens *during and around* the search, rather than before it begins. The MERIT system [16], shown in on a much more flexible conversation model that took place during a search, with the system responding in some form at each stage of the dialogue.

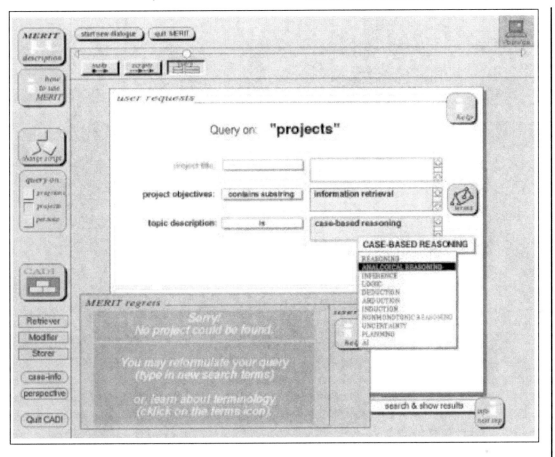

Figure 3.2: An advanced IIR SUI, called Merit, that responds based on a complex model of human conversations; taken from [16].

3.1.2 BROWSING

Another early type of system, given the command line technology available at the time, was represented by browsing SUIs. Similar to the initial dialogue-based systems, browsing systems like the 1979 BROWSE-NET [147] scan through databases. Although it was maybe not entirely obvious at the time, BROWSE-NET was supporting Nielsen's recognition heuristic more than the dialogue-based systems that were determined by sequential recall and text-based input.

We now see these browsing style systems appear in many SUIs, although in 1983, Geller and Lesk showed that people 'browsed' less on the early online newsgroups [61]. Gellar and Lesk hypothesised that this may have been because people often knew more about what was in a fixed dataset than in an oft-changing web collection. Despite this hypothesis, we later saw the rise of

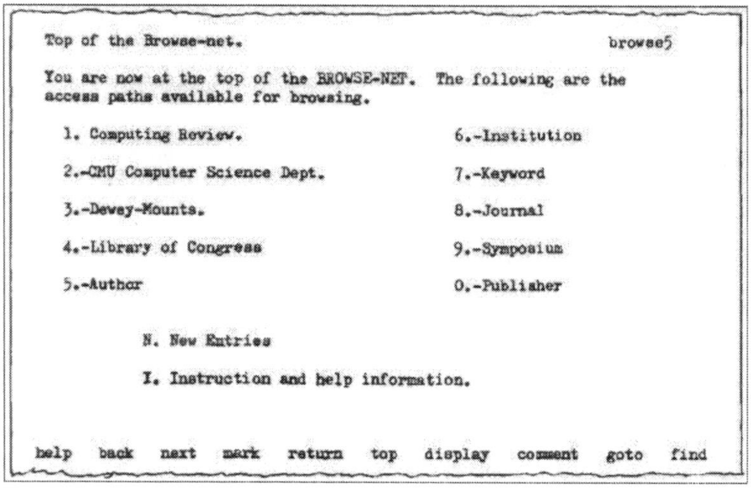

```
Top of the Browse-net.                                    browse5

You are now at the top of the BROWSE-NET.  The following are the
access paths available for browsing.

  1. Computing Review.                  6.-Institution

  2.-CMU Computer Science Dept.         7.-Keyword

  3.-Dewey-Mounts.                      8.-Journal

  4.-Library of Congress               9.-Symposium

  5.-Author                             0.-Publisher


        N. New Entries

        I. Instruction and help information.


help   back   next   mark   return   top   display   comment   goto   find
```

Figure 3.3: An early browsing interface for databases that provides options for different ways of accessing the documents; taken from [147].

website directories, like Yahoo! Directory[1] (Figure 3.4), LookSmart, and the recently retired Google Directory. Directories, while still available[2], were extremely popular in the mid 1990s, competing with early search engines like Lycos, AltaVista, and Excite. While these early search engines tried to increase accuracy and speed, 'Surfing the Web' using well-structured directories like Yahoo! Directory was a popular past time, especially when searchers could 'see what was new' using a service from Yahoo! called 'The Spark'[3]. When Google arrived, however, its accuracy and speed at searching the growing Web [25] meant that people quickly switched to search engines. We do, however, still see browsing interfaces appear within specific websites, where the website owners know more about the collection, and it is easy to define what categories searchers may want to browse by. Such conditions are discussed further in relation to 'Faceted Browsing' in Section 4.

3.1.3 FORM FILLING

As SUIs became more directly interactive, with the onset of commercially available Graphical User Interfaces (GUIs) in the late 1980s[4], the common paradigm we see today of 'Form Filling' became more popular. Although text-entry command-line systems could be turned into spatial systems, the popular onset of GUIs converted conversational response SUIs, which took input sequentially, forward by providing all the data entry fields spatially in one view. Even recent HCI research [168]

---

[1] http://dir.yahoo.com

[2] http://www.dmoz.org/

[3] http://dir.yahoo.com/new/

[4] GUIs were increasingly available in research and development in the late 1970s and early 1980s, but the first Mac and Windows GUI-based personal computers grew in sales towards the late 1980s.

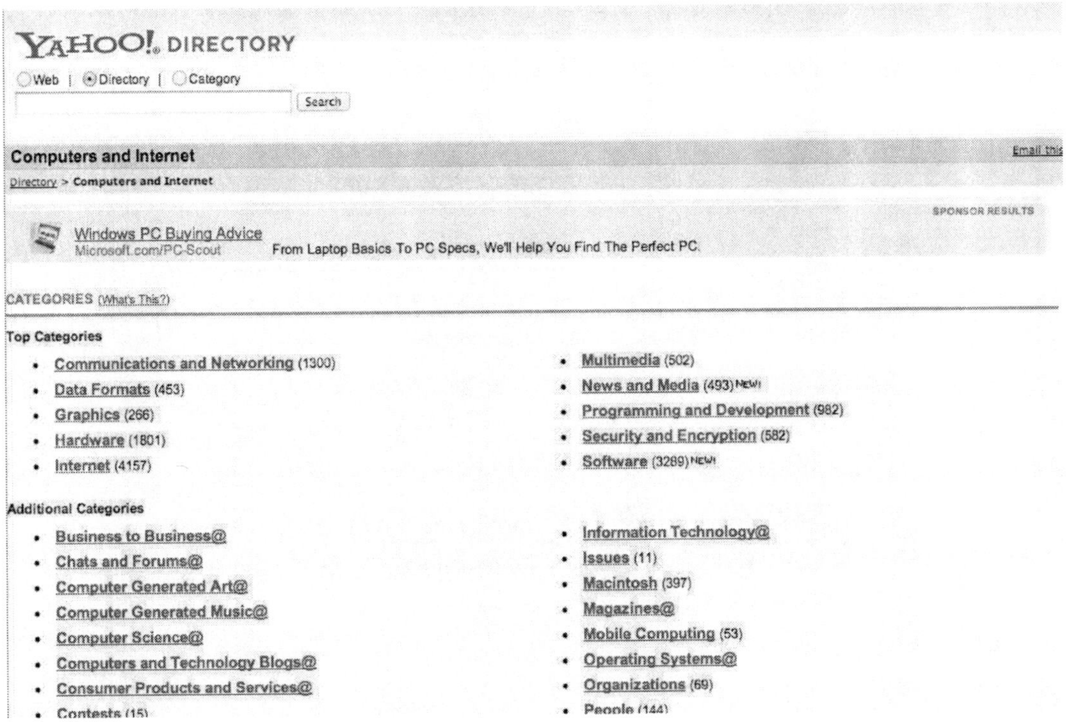

Figure 3.4: Yahoo! Directory, like other web directories, was extremely popular until the efficiency and scale of search engines like Google improved.

and cognitive science literature [145] emphasise the benefits of spatial rather than temporal inter-action. Although 'Form Filling' includes normal keyword searching, this technique allowed systems to present all the fields that could be individually searched in a way that we now commonly call an 'Advanced Search'. Form filling did, however, allow researchers to develop early template-oriented SUIs that were customised to a particular dataset. The EUROMATH system, shown in Figure 3.5 designed by McApline and Ingwersen [123], has a custom form highlighting all the fields that can be searched individually or in combination.

3.1.4 BOOLEAN SEARCHING

One advance in the algorithmic technologies was to process Boolean queries, so that we could ask for information about 'Kings OR Queens', and get a more comprehensive set about, in this case, Monarchs. This technological advance was made before the majority of SUI developments, as can be seen in Figure 3.5. The advent of GUIs, however, provided an opportunity to help people construct Boolean queries more easily and visually. The STARS system [7], shown in Figure 3.6, allowed

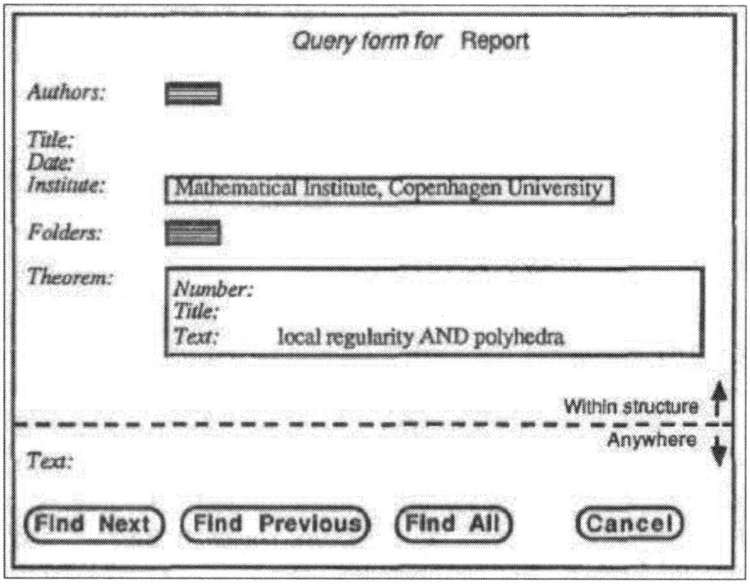

Figure 3.5: The EUROMATH interface had custom forms for each type of record in the system; taken from [123].

searchers to organise their query in a 2D space, where horizontal space represented 'AND' joins, and anything aligned vertically were 'OR' joins.

Like all these early ideas, Boolean searching is still prevalent in our modern SUIs. Figure 3.6 shows a Boolean query being performed in Google, using more layman-language operators; the '-' before a search term is equivalent to a Boolean NOT, in this case. Most modern query algorithms now use probabilistic methods to rank results [88], where the equivalent of ANDing all the terms is prioritised, but results without one or more of these terms can still be returned if the page is highly relevant to the keywords. These systems, however, still typically allow Boolean queries to be performed too.

3.1.5   INFORMATIONAL ADVANCES

The initial advances in IR were typically made in technological improvements. Referring back to the level model shown in Figure 2.1, improvements were in the lowest two layers, where better algorithms allowed the same SUI to accept less rigid and more powerful queries (like the Boolean queries). Other algorithmic advances allowed queries to be enriched by the common words that were in the top results (e.g., Automatic Query Expansion [26]). Consequently, these SUI advances made in the early days related mainly to the *Input* SUI features, with the exception that the structure of some of advances (like the Browsing and Form Filling) provided information about the structure of the

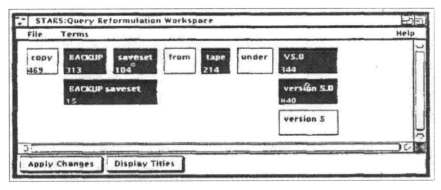

Figure 3.6: The query builder in the STARS interface allowed searchers to arrange query terms in a 2D space to build a Boolean query. Items aligned vertically are ORed and items aligned horizontally are ANDed; taken from [7].

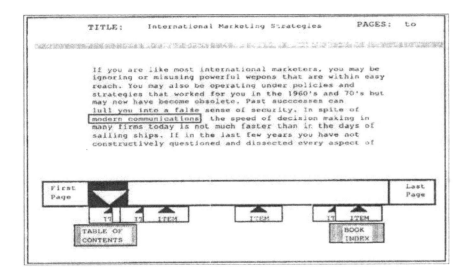

Figure 3.7: This demonstration system highlighted, in a horizontal overview, where a search term appeared in a document; taken from [193].

data, making them also contribute to the *Informational* SUI features. Other *Informational* advances were to highlight in a result where it matched the query. The visualisation shown in Figure 3.7 allowed searchers to scroll through the full text of a book, where the horizontal bar at the bottom indicated where in that book the search terms appear [193]. We now see this type of *Informational* feature in current systems too, such as the indicators sometimes embedded in the scrollbars of some software (shown in Figure 3.8) or in the HotMap SUI (shown later in Figure 4.26), which has an overview of how related each of the first 100 results is to the query.

Figure 3.8: When searching for terms inside a web page, the Rockmelt[5] browser highlights search terms both on screen, and off-screen using indicators in the scrollbar.

The onset of GUIs meant that SUIs became more interactive, with Pejtersen's fiction browser [150] presenting an explorable-world view of a bookshop, as shown in Figure 3.9. The research and SUIs discussed in this section, however, are not yet representing what would now be called interactive IR or exploratory searching, which we consider to be a more on-going period of interaction over multiple searches to reach a goal, rather than the single search. In the next subsection, we describe SUI design issues that relate to our current notion of IIR, beginning with a key study of a SUI that showed that interaction provided significant benefits to searchers.

---

[5]http://www.rockmelt.com

Figure 3.9: Pejetersen's fiction bookshop allowed searchers to browse the bookshop using different strategies, where the figures shown are engaging in each strategy; image from [150].

3.2 THE ONSET OF MODERN SUIS

The onset of modern interactive SUIs began around of the time that we first saw web search engines like AltaVista[6], but before Google was launched. One of the first studies to demonstrate that there were significant and specific benefits to IIR, where searchers actively engage in refining and submitting subsequent queries, was provided by Koenemann and Belkin in 1996 [98]. Using a popular, at the time, query engine called INQUERY, Koenemann and Belkin built the RU-INQUERY SUI, shown in Figure 3.10 (b). Searchers could submit a query in the search box at the top-left, and see a scrollable list of results on the right-hand side (showing 5 at any one time). The current query was then displayed in the box underneath the search box. The full document of any selected result was displayed beneath the results on the right.

The experiment was built to leverage Relevance Feedback [162], which is an iterative process that lets searchers explicitly, or implicitly with clicks, judge results as being relevant or not. In principle, results that are determined as relevant are used to extract more keywords to augment a query, thus allowing the system to find more results that are similar to the keywords from both the query and example relevant documents. Salton and Buckley compared several algorithms for implementing

[6] http://www.altavista.com

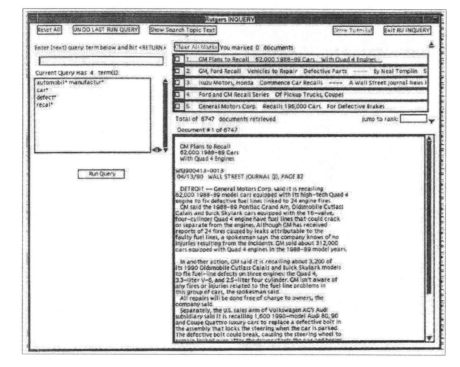

Figure 3.10: The RU-INQUERY interface was studied with hidden, visible, and interactive relevance feedback terms, where interactive provided the most effective support for searchers; taken from [98].

relevance feedback [165], but the work at this time had simply proved that more relevant results could be returned with knowledge of relevant results for a given query. Indeed, most systems that used Relevance Feedback did so as a background augmentation to the algorithm. Koenemann and Belkin, however, were amongst the first to show that actual interaction with Relevance Feedback terms, rather than simply providing judgements, could help searchers make significant gains in task performance.

In a very explicit form of Relevance Feedback, RU-INQUIRY took the key terms from the results marked as 'relevant' and added them to the search. Searchers marked documents as 'relevant' by ticking a checkbox shown next to each result. To demonstrate the benefits of explicit interaction in IR, rather than simply improving algorithms with relevance feedback data, they developed three alternative versions:

1. **Opaque** provided the typical Relevance Feedback experience that was common at the time, where terms from the selected *relevant* documents were added, but there was nothing in the SUI to say what those additional terms were.

2. **Transparent** provided a similar experience to the **Opaque** version, except that the added terms were made visible in the 'Current Query' box.

3. **Penetrable** allowed the participants to choose additional terms from the relevant documents. The keywords associated with the *relevant* documents were listed in a separate box below the 'Current Query' box (as shown in Figure 3.10 (a), and could be added to the current query box manually.

These three experimental conditions were compared in a user study to the original version that did not have relevance feedback. While all three experimental versions improved the number of relevant documents found and reduced the searching time, the most interactive **penetrable** version provided statistically significant accuracy improvements and did not significantly increase the time involved in searching. While previous work had already shown that Relevance Feedback and other query expansion methods could improve the precision of results, this study showed that engaging and interacting with queries and results brought significant benefits to the searcher.

These clear results motivated many of the SUIs that we see today despite the fact that explicit relevant feedback has itself become less popular; it became commonly accepted that searchers rarely stop to provide relevance judgements, despite the potential gains provided by doing so. Back et al. concluded that the process of rating the relevance of documents involved too much of a cognitive switch during search, and so detracts from the act of finding the sought information [8]. We still, however, see alternative forms of interaction and feedback in IR, described in the next section. Encouraging searchers to submit longer queries, using Interactive Query Expansion, for example, was originally shown to provide algorithmic benefits (e.g. [72]), but experiments with searchers in the late 1990s began to show how this would be used in practice by searchers. Magennis, for example, showed that experienced searchers, who chose good query suggestions, could get a small but significant gain in search quality, but that inexperienced searchers often chose ineffective query suggestions [119]. Ruthven later showed that systems were more likely to identify good query expansions than human searchers, indicating that to help searchers attain these significant gains, systems should first select valuable and effective query suggestions. Research by Kelly et al., however, concluded that searchers *were* able to identify useful query suggestions, with information about the 'popularity of suggestions' having little effect on how they were used [96].

3.3 SUMMARY

In terms of the framework of features types described in Section 1.3, the ground-breaking studies in the mid to late 1990s showed the initial value of having *Control* SUI features to help modify and manipulate a search. This change highlights a transition from SUIs that simply provide *Input* to, or have *Informational* displays about, an efficient and effective search system (Section 3.1), to SUIs where the *Control* and *Personalisation* makes the system efficient and effective. Sections 4 and 5 now consider many novel designs and features that innovate in all 4 of these aspects of SUI designs.

CHAPTER 4

Modern Search User Interfaces

This chapter covers many of the more modern advances in SUI designs, and is structured according to the framework described in Section 1.3. The chapter begins by discussing *Input* features, before moving on to *Control, Informational* and *Personalisable* features. Additionally, some specific SUI design recommendations are highlighted throughout this chapter.

4.1 INPUT FEATURES

While there have been many technical advances in processing searcher queries and matching them against documents, the plain white search box has remained pleasingly simple. While there have been some variations in how we enter information into a search box, the alternative to searchers manually generating possible words for queries, is to present metadata that can be selected and utilised. This section begins with the minor advances in the design of a search box, before discussing the different ways in which metadata can be helpfully presented to searchers as an Input feature.

4.1.1 THE SEARCH BOX

The search box pervades SUIs and searchers can feel at a loss when they do not have a small white text field to spill their search terms into. The search box has many advantages: it is extremely flexible (assuming the technology behind it is well made), and it uses the searchers language (one of Nielsen's heuristics). The searcher can be as generic or specific as they like, especially if they know the special command operators. So how, if at all, have search boxes advanced over time? As well as being primarily used for an *Input* feature, the Search Box can — and should — be used as an *Informational* feature. When not being used to enter keywords, the search box should be informing the searcher of what is currently being searched for. Using the search box to display the current query has at least two benefits. First, whether a search system corrects a spelling mistake in a query, or allows it to continue, the searcher may always want to correlate what they are seeing with exactly what is being searched for. Bates noted that comparing results with the current search term is a common search-monitoring tactic [13]. Displaying the current query makes the search box *Informational*, as well as for *Input*. Second, should searchers wish to change their search in any way, it would be painstaking to require them to step backwards for one or more pages in order to edit their query, or even submit a new one. Efficiency, or saving searchers' time, is one of the common design heuristics. Keeping the search box, and the current query, visible at all times makes it easy for searchers to use it as a *Control* feature too.

> **Recommendation**
>
> - Keep the search box and the current query clearly visible for the searcher at all times.

Auto Complete

One of the key problems with a search box is that it is easy for a searcher to do something that does not retrieve the results they want. Nielsen refers to this problem in his Control and Freedom heuristic and his Error Prevention heuristic. This problem was especially important when search systems took a long time to respond, but is still true when the searcher fails to get the results they are looking for straight away. Auto-complete helps overcome these problems by giving people guidance towards queries that are likely to work. Again, in terms of Nielsen's heuristics, this allows the search box conform to the recognition over recall principle too. Given that auto-complete also provides information to the searcher as they query, auto-complete helps make the search box a better *Informational* feature as well as an *Input* feature. Further, this applies the findings of Interactive Query Expansion (discussed briefly in Section 3.2), to support searchers before they have even performed their first search. Auto-complete can be rich with context, with the Apple website providing images, short descriptions and even prices, as can be seen in Figure 4.1 (a). Further, auto-complete can be *Personalisable*, as with Google in Figure 4.1 (b), showing searchers, with a Google account, queries that they have used before in their search history[1].

> **Recommendation**
>
> - Help searchers to create useful queries whenever possible.

Operators and Advanced Keywords

The keyword search box itself has only really received minor visual changes, with some suggesting this may affect the number of words people put in their query [139]. Regardless, studies indicate that searchers submit between 2 and 3 word queries [84, 93], and around 10% of searchers use special operators to block certain words or match explicit phrases. Advanced Search boxes, when implemented well can help guide people towards providing more explicit queries in the search box.

---

[1]The presence of search history in the auto-complete appears to vary according to the technology, such as the browser, being used.

(a) Apple – shows lots of contextual informa-   (b) Google – prioritising previous searches.
tion and multimedia.

Figure 4.1: Examples of AutoComplete.

The majority of the fields in Google's advanced search box can be translated to special operators in the normal query box. Consequently, when the results are displayed, the full advanced search form does not also have to be displayed (helping maintain Nielsen's consistency heuristic for the design of SERPs). Further, the expert searchers can use shortcuts (another of Nielsen's heuristics) by using the operators instead of the advanced search form.

Query-by-Example

There is a range of searching systems that take example results as the *Input*. One example commonly seen in SERPs is a 'More Like This' button, which returns pages that are related to a specific page. Google's image search also provides a 'Similar Images' button, which returns images that appear to be the same, in terms of colour and layout. While these could be seen as *Control* examples (modifying an initial search), the Retrievr prototype SUI (Figure 4.2) lets a searcher sketch a picture and returns similar pictures. Similarly, services like Shazam[2], let searchers record audio on their phone and then try to find the song that is being played. Shazam and Retrievr are examples that are explicitly query-by-example *Input* features, while others can be seen as *Input* and/or *Control*.

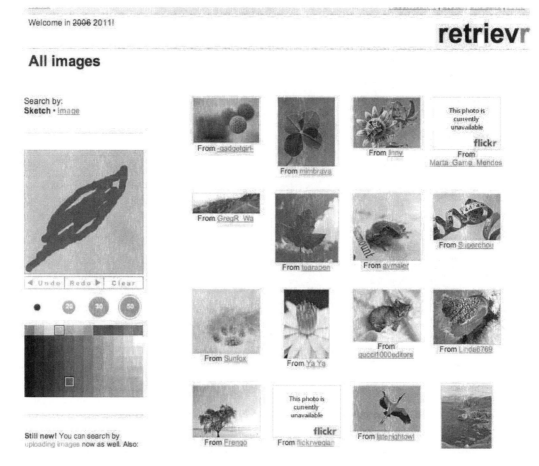

Figure 4.2: Retriever is a prototype system that lets searchers sketch what they are looking for; called query-by-example.

Scholars have been investigating query-by-example systems for some time, with the concept of query-by-humming being introduced for music collections by Ghlas et al. in 1995 [63]. A study of the CueFlik prototype has shown that searchers can find better images by providing regular and frequent feedback about good and bad images [4]. Yeh et al. also showed that screenshots of error messages and configuration screens were particularly useful for finding technical support documents [218]. One indicator of the maturity of this interaction is that Google now allows searchers to upload images for their first query, rather than using words as descriptors. Further to make the UX more intuitive, Google lets searchers drag and drop images into the search area, as shown in Figure 4.3. Beyond this, Google also allows searchers to submit web searches by speaking a query into their microphone; a search experience afforded by both advances in speech-to-text algorithms and in Web UI technologies like HTML5[3].

Figure 4.3: Google Image Search allows searchers to upload, link to, or simply drag and drop images into their search box, allowing searchers to query by example.

4.1.2 ADDING METADATA

The metadata that is often provided to searchers in SUIs can vary in both depth and complexity. Further, metadata can serve many purposes, providing *Input, Control,* and *Informational* support, and could have been included in several of the sections. Some metadata, for example, is only shown after an initial search, which would make it primarily a *Control* feature. Similarly, lots of metadata is provided as simply for its *informational* value. Metadata is described here, however, as in many of the systems described below, SUIs allow searchers to use it from the very beginning as a form of primary *Input.*

Website Directories, including the Yahoo! Directory (Figure 3.4), are a clear example of using metadata as an *input,* as it provides higher-level categories to help searchers externalise what they are looking for. Several studies of SUI prototypes, including the Superbook, Cha-Cha, and WebTOC, have shown that categorising results in SUIs can help searchers to find results, for example, 25% faster and more accurately [33, 50, 52, 135]. Further, Superbook's use of categories was shown to

---

[3]http://www.w3.org/TR/html5/–the W3C's specification for HTML5.

improve learning, as measured by short open-book essays. Work by Gwizdka indicates that such category labels for each result are particularly helpful during the stages of query formulation and result examination [66].

> **Recommendation**
>
> - Make it clear how results relate to metadata in your system, to help searchers to judge the results and make sense of the whole collection.

Hierarchical categories are still frequently used within many modern SUIs. eBay and Amazon, for example, provide searchers with a hierarchy of categories so that they can first define what *type* of object they are looking for. eBay then lets searchers filter the results using additional metadata that is specific to that *type* of object.

As noted above, the presentation and use of metadata in SUIs can be very hard to delineate in its multiple contributions to *Input, Control, Informational*, and *Personalisable*. Indeed, well-designed use of metadata can serve as a feature in each of these four groups. Presented on the front page of a SUI, categories can, for example, allow the searcher to *Input* their query by browsing. If a searcher can use metadata to filter a keyword search, or to make sub-category choices, then metadata can quickly become a *Control* feature. Further, if search results are accompanied by how they are categorised, then metadata can become an *Informational* feature. Research by Drori and Alon indicated that showing results along with their categorisations (a) reduced search time, (b) increased result accuracy, and (c) preferred by participants [49]. Finally, it's not beyond the realm of possibility to highlight popular or previously used category-options to make them *Personalisable* too.

Clusters

One challenge for providing category-based metadata, especially for the whole web, is to categorise all the results. Another approach, using clustering algorithms in the back end, is to cluster results by key topics in their content. There are many algorithmic advances for clustering that are too detailed to be discussed in this book (Zhao and Karypis compared 15 hierarchical clustering algorithms in 2002 [220]). An early system called Scatter/Gather (Figure 4.4) showed that clustering results into different major groups, as an *Informational* feature, helped searchers to be more efficient and finding results [80]. Leuken et al. have also shown that clustering image results can be used effectively to diversify search results, avoiding returning the same image many times over [113].

A system originally called Clusty[4] (Figure 4.5) took a different approach, and used the clusters within results primarily as a *Control* feature. Rather than visually clustering search results, the

[4]http://www.clusty.com

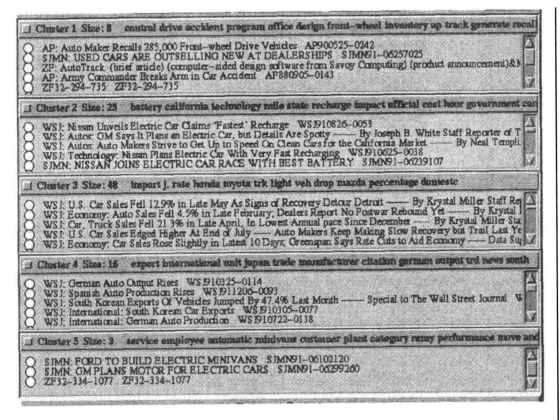

Figure 4.4: The Scatter/Gather interface clustered results into major groups, highlighting for each cluster: how big it was, and common words within it, image from [80].

hierarchically organised clusters were presented on the left-side of the screen and could be used as a filter. Despite some studies showing evidence that clusters help searchers to search (e.g., [197]), research has suggested that well-designed carefully-planned metadata is always better for SUIs than automatically generated annotations [77]. Well-designed metadata is so called, however, if it 'speaks in the searchers language,' and avoids technical jargon (Nielsen's language heuristic).

Faceted Metadata

More recently, it has been popular to categorise results in multiple different dimensions, so that searchers can express several constraints over one result set. Research has shown that, compared to keyword search, faceted systems can improve search experiences in more open-ended or subjective tasks (where no single right answer is available) [183]. Studies of a system called Dyna-Cat, for example, which automatically generated facets for medical search results, indicated that participants

Recommendation

- Carefully curated metadata is better than automatically generated, but both are better than no metadata at all.

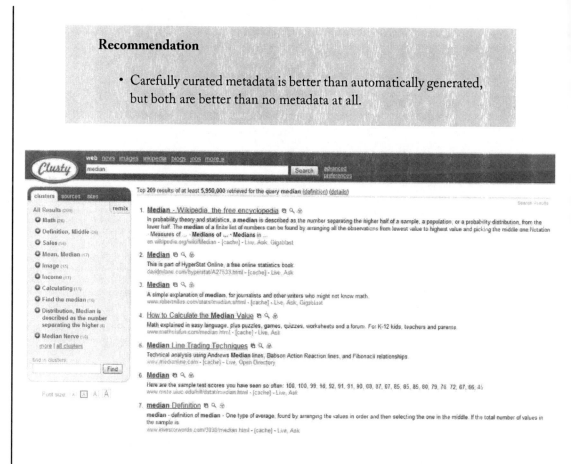

Figure 4.5: Clusty creates automatic hierarchical categories based on the results that are returned.

completed fact-finding tasks 50% faster than with a standard SERP [153]. We now see facets embedded in many SUIs. Epicurious[5], shown in Figure 4.6 for example, allows searchers to describe recipes that they would like by several types of categories (called facets), including cuisine, course, ingredient, and preparation method.

There are many different variations in the design of facets. The number of options in each facet-category can be long, such as the ingredient facet in Epicurious. To overcome long lists, eBay[6] shows the most popular by default and shows the rest if asked. SUIs may show all the categories at once, like eBay, or show one at a time (by minimising the others) like Epicurious. Each facet may be hierarchical, as demonstrated by the research prototype: Flamenco[7] [217]. A faceted browser may

---

[5]http://www.epicurious.com/
[6]http://www.ebay.com
[7]http://flamenco.berkeley.edu/

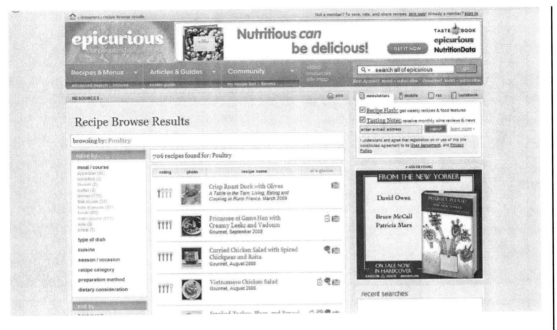

Figure 4.6: Epicurious provides several different categories (called facets) that can be used in combination to define a query.

allow multiple selection (i.e., to view cameras from two companies, rather than just one) as in eBay. Faceted browsers can hide used facets like Epicurious and Flamenco, or show the decisions that a searcher has made, as in the more experimental faceted browsers called Exhibit[8]. Exhibit leaves a facet present in the user interface immediately after a decision is made, in case the searcher wants to change their mind. Further, a more advanced faceted browser called mSpace[9], shown in Figure 4.7, leaves the facets in view permanently, so that a trail of the searcher's decisions, in the context of the options, can be seen at all times [169]. The visualisation of a trail in mSpace is facilitated by the horizontal alignment of facets, similar to iTunes, where facets only filter facets placed to the right.

The common alternative to mSpace's spatial trail approach is to place the selections in a breadcrumb (discussed further below), and have to remove that decision before being able to (a) see the alternatives and (b) change it. SUI designers should think about whether searchers are more likely to explore and change decisions, or simply try to express their criteria and rarely backtrack. mSpace supports changing and backtracking more easily, while Flamenco and many other faceted SUIs aim primarily to support new filtering decisions by allocating them more space.

[8]http://www.simile-widgets.org/exhibit/
[9]http://mspace.fm

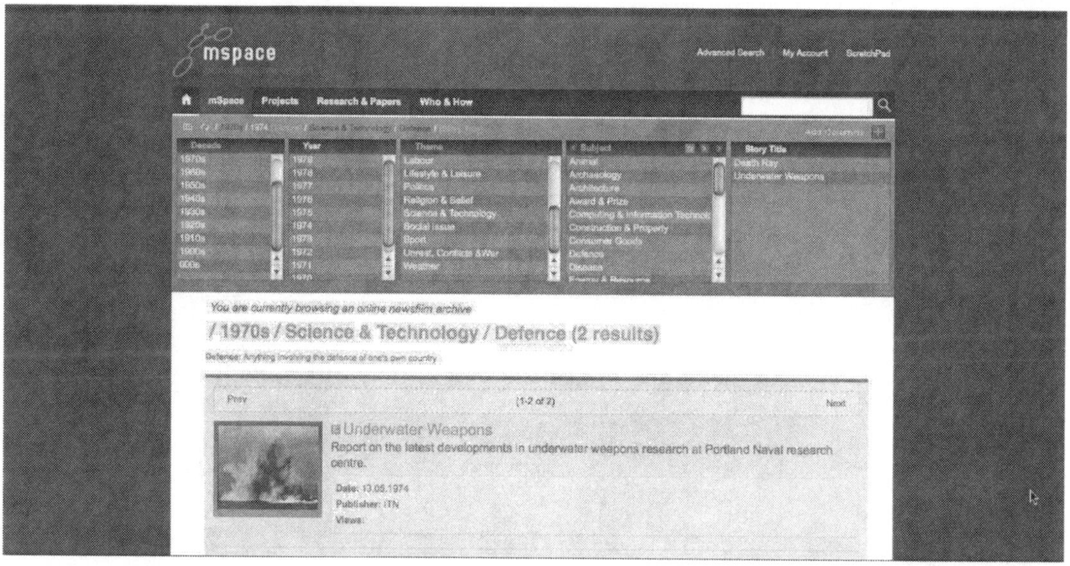

Figure 4.7: mSpace provides an advanced faceted SUI where the order of facets implies importance and gaps from left to right are highlighted. This figure shows that the two clips in the far right column are from the 1975 and 1974; which would not normally be conveyed in faceted SUIs.

Recommendation

- If appropriate, support searchers in reviewing their decisions and their options quickly and easily.

Facets can become even more *Informational*, beyond displaying the facets on the screen, by displaying how results are related to a search result. The Relation Browser visualises how many of the results are related to every item in each facet using a barchart style visualisation [219]. Alternatively, the SERVICE prototype (Figure 4.8) shows which items in each facet a result is related to, as searchers hover over each result [106, 107]. Similarly, when a result is selected in mSpace, each facet is used to visualise related metadata, making them even more *informational*. mSpace also highlights related selections in unused facets to the left of any selection within the facets [208]. Conversely, systems like ASOS and the BBC's podcast explorer[10] grey-out faceted items that are not related to the current search. This allows searchers to know which categories exist (Nielsen's consistency

[10]http://www.bbc.co.uk/podcasts

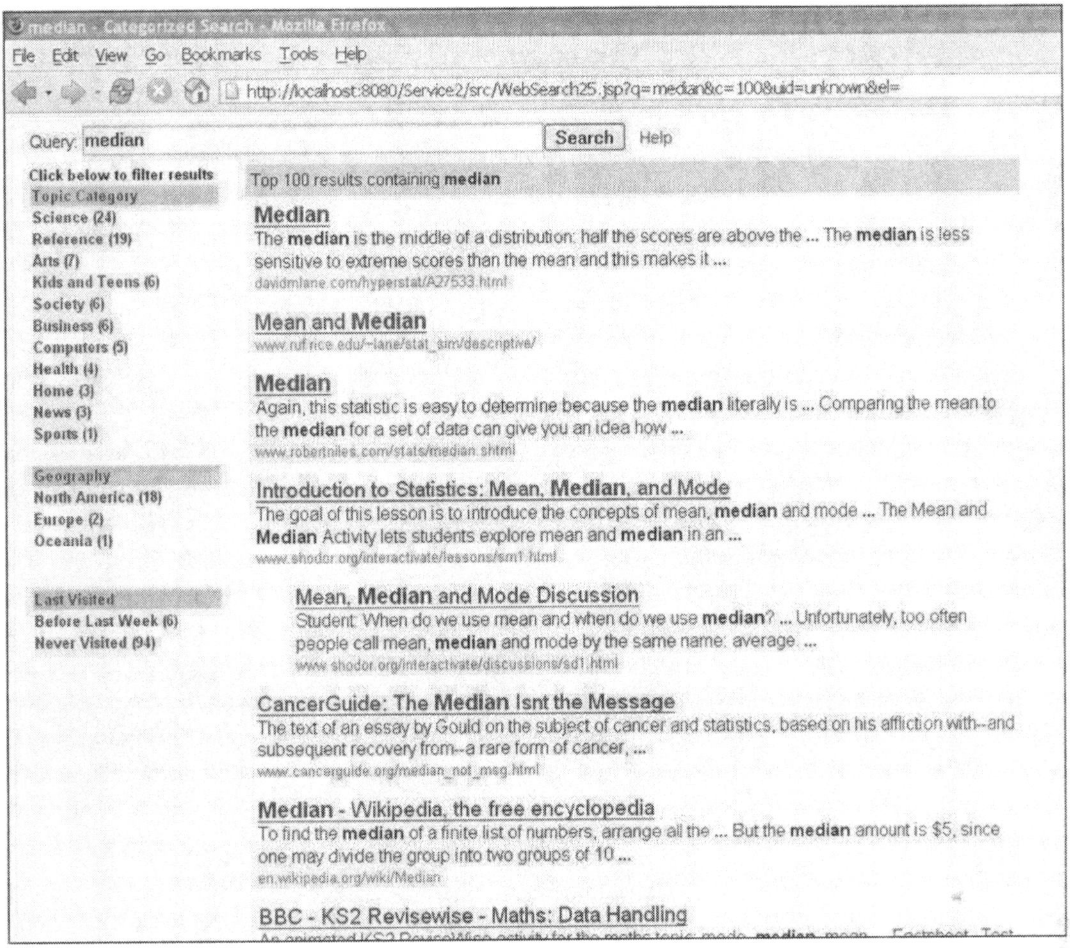

Figure 4.8: SERVICE provides faceted browsing over generic web search results.

heuristic), while guiding them away from unproductive selections (another of Nielsen's heuristics). Further, the BBC podcast website allows searchers to select greyed-out results, overriding their current search; this further adheres to Nielsen's flexibility heuristic. By using the left-to-right order of facets, horizontal faceted browsers, like iTunes and mSpace, can also visualise more subtle details about how facets are related. iTunes, unlike the approach taken by Epicurious, can display all the artists related to one Genre, while also showing all the albums by one artist. mSpace furthers this horizontal approach by allowing facets to be reordered, removed, and replaced with alternative facets. These advantages created by using the horizontal faceted approach also make facets more *Informational*.

Using highlights within metadata that are stable (rather than determined by mouse-hovering) can also make facets more powerful for *Control*, by guiding searchers towards selections. Further, allowing searchers to control the order of the facets in mSpace means that searchers can state that price is more important than quality, for example, by putting the price facet to the left of the quality facet.

A more in-depth and critical analysis of how multiple faceted categories are designed has been provided by Clarkson and colleagues [36]. There are also books on faceted search [196] and design recommendations for the more common forms of faceted SUIs on the web [75].

Despite examples like SERVICE attempting to calculate faceted metadata for the whole web, most search engines do not provide faceted filters on their main web search service. Google, however, does provide facets in their product search[11]. Faceted metadata categories are typically used within fixed collections of results, such as within one website (typically called vertical search), as there must be common attributes across all the data to categorise them effectively. This is an example of where scenario of use and the type of metadata both inform the design of a SUI. In the narrower information space of products, there are more common factors like price and retailer reputation that apply to all of the results. It might also be noted that, until 2011, Google product provided facets at the bottom of the search, perhaps because it treats facets as more of a *Control* filter than a proactive *Input* method. For reasons that appear to be more aligned with maintaining a unified and consistent UX (as per Nielsen's consistency heuristic), Google product search has adopted their facets into the left-most column on the screen, which is now a panel used almost entirely for *Control* features on the majority of their services. The placement of *Control* features on the left was primarily led by Bing's UX changes in 2009, who also use this space for facets in their product search: Ciao!

Social Metadata

Since the rise of more social websites like Facebook, Twitter, and Google+, it has been popular to use socially generated metadata, such as tags, to help searchers *Input* their query. Tag clouds are now familiar in many SUIs, with services like Flickr allowing searchers to explore pictures by popular tags[12]. Again, by providing related tags, when viewing results relating to a current tag, the searcher can also refine (*Control*) or change their search. Gwizdka studied a prototype that created per-query tag cloud overviews, as well as smaller tag clouds per search result, and showed that they were especially helpful for people with strong verbal cognitive skills [67]. Kuo et al. studied the inclusion of tag clouds in the PubMed, indicating that they were especially useful for simple search tasks, but not for discovering relationships between concepts [109].

Early findings from Wilson and Wilson indicate that merely the presence of a tag cloud in a SUI can help searchers make sense of results and information, rather than simply being used for issuing new queries [206]. A research prototype called Mr Taggy[13] (Figure 4.9), however, allows

---

[11]http://www.google.com/products
[12]http://www.flickr.com/photos/tags/
[13]http://mrtaggy.com/

searchers to search by different types of tags, by separating adjectives and nouns. Studies of MrTaggy suggest that the searchers explored more and learned more when tag clouds are available [91].

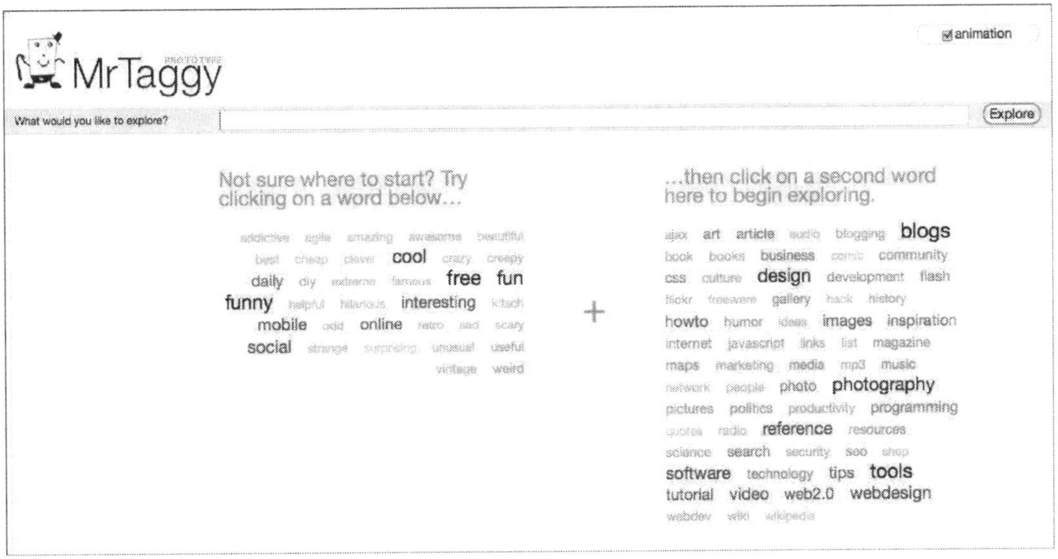

Figure 4.9: The MrTaggy prototype provides two tag clouds (adjectives and nouns), collected from Delicious, to help searchers explore the web.

Other social metadata can be used, but is often *Informational*, such as how other searchers have rated a result, or *Personalisable*, such as the searching actions of people that a searcher knows. As stated above, much of this metadata can be for *Control* as well as *Input*.

4.2   CONTROL FEATURES

Control features can be considered particularly important from an IIR perspective, as they facilitate and control the ways in which the search continues and thus makes the experience interactive. Research has emphasized the importance of good *Control* features, showing that the right support can significantly enhance an IIR task, but the wrong support can distract or slow down searchers [47]. The section above has already touched on how *Input* features can be reused for refining searches. Categories, for example, can be used to narrow down the search results produced by a query.

It is a good SUI design principle, as obeyed by Epicurious and most search systems, to always return results after the first action, rather than also requiring a *Control* action. There is rarely a good reason to ask the searcher to submit a second filter before displaying results that match the first, especially if the first interaction alone may happen to return the desired result. This principle has been taken to the extreme by Google Instant, which starts showing search results (not just query suggestions) after the first character of a query has been typed.

Recommendation

- Always return results based on the first interaction, as subsequent interactions may never be needed.

4.2.1   INTERACTIVE QUERY CHANGES

One of the key ways that we continue to interact with a query is to alter and refine a search, known as Interactive Query Expansion (IQE). The aim of IQE is to suggest additional, or replacement, words to the searcher that might help the system return more precise results. If, for example, a searcher searches for 'Queens of the stone age,' SUIs often return a series of extensions to that query that might further define what the searcher is looking for, as shown in Figure 4.10.

The technological implementation in the backend varies, as briefly noted at the end of Section 3.2, but the choice of refinements and the position in the SUI should be determined by the expected searchers and the types of tasks. Google, while focusing on the precision of the search results, presents IQEs at the bottom of the SERP, perhaps assuming that searchers will only want to *Control* their query if they don't find what they want in the first 10 results. Further, Google only presents IQEs if there is a notable range of smaller types of target with commonly used queries. In addition to these expansions, Google also presents alternative-direction suggestions (to change focus rather than to narrow) on their left-hand *Control* panel, under the title of 'Something Different.' Bing[14], in comparison, defines itself as a 'Discovery Engine' and provides a collection of both refinements (to narrow) and alternative searches (to change focus) on the left side of all SERPs, as shown in Figure 4.10 (a). Similarly, as an example of a vertical search system, Amazon[15] provides a combination of refinements and alternatives to a search, as shown in Figure 4.10 (c). As an online retailer, Amazon's motivation for placing refinements and alternatives at the top of the page may be to encourage customers to discover and buy more products.

4.2.2   CORRECTIONS

One good rule for SUIs is to never let the searcher reach dead-ends where the interaction can only stop, and so another common form of *Control* feature helps correct and notify searchers when they are heading down a fruitless path. In purposefully submitting a query with two errors, such as 'Quens of the Stonage' for example, services often suggest query corrections, or simply auto-correct the search. For this example, Amazon states clearly that no results were found, provides the spelling correction alternative, and returns example results for that spelling correction. Bing and Yahoo!, as

---

[14]http://www.bing.com
[15]http://www.amazon.com

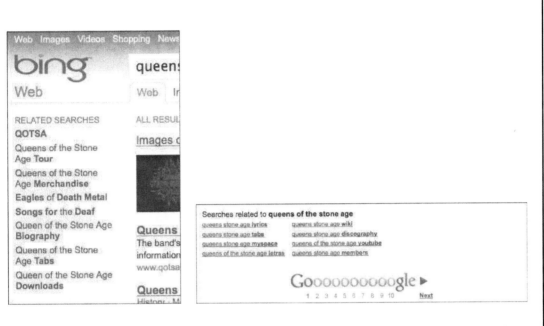

(a) Bing – IQEs on left of SERP

(b) Google – IQEs for some searches at the bottom of the results.

(c) Amazon provides fewer IQEs above the results.

Figure 4.10: Interactive Query Suggestions (both refinements and alternatives) are a familiar sight in most search systems.

(a) Bing (shown here) and Google often auto-corrects errors, but provides a link to results on the exact query, if it is confident that the corrected version is more likely.

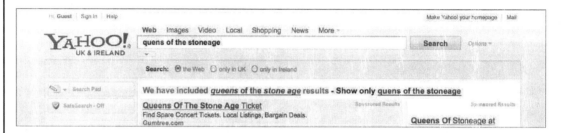

(b) Yahoo! currently only auto-corrects errors, but provides a link to results on the exact query.

(c) Amazon says it cannot find results for an incorrect query, but provides a link and example results for a corrected version.

Figure 4.11: Correcting errors in searcher queries to avoid dead ends.

shown in Figure 4.11, however, inform the searcher that they are auto-correcting the errors, but offer the searcher to search formally for the original version. Google and Bing can also vary on how they respond with corrections, likely basing the decision on the confidence of the two options. When the most-likely query is less clear-cut, as with searching for 'Amzon,' both appear provide a simple 'Did you mean:.' When the decision is more obvious, however, they both side with the significantly more common query, as Bing does in Figure 4.11 (a).

Recommendations

- Never let searchers reach a dead end, where they have to go back or start over.

- Help searchers avoid mistakes wherever possible, but do not force that help upon them.

4.2.3 SORTING

One method of helping searchers stay in control of what they are looking for is to allow them to decide how they want results to be ordered. Web search engines typically order results by how relevant they are to a query, and to allow searchers to filter them. It is common in vertical domains like online retail or document repositories, however, to be able to change results so that they are ordered by factors such as price, and either by most expensive to least expensive, or visa-versa. Figure 4.12 (a)-(c), for example, shows a range of ways in which products can be reordered. Scan.co.uk and iTunes (Figure 4.12 (d) and (e)), however, take an approach that is more commonly seen in tabular views, where any column of metadata can be used to order results. Choosing a column orders the results by the metadata in that column. Re-choosing a column typically reverse-orders them.

Recommendation

- Give searchers control over the way results are ordered.

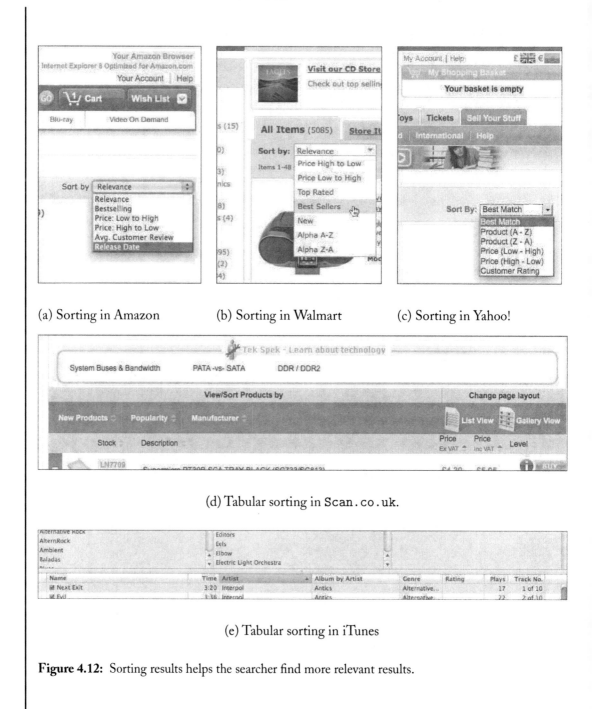

(a) Sorting in Amazon   (b) Sorting in Walmart   (c) Sorting in Yahoo!

(d) Tabular sorting in `Scan.co.uk`.

(e) Tabular sorting in iTunes

Figure 4.12: Sorting results helps the searcher find more relevant results.

4.2.4   FILTERS

Although not too dissimilar from using metadata-based *Input* methods as a *Control* feature, as discussed above, SUIs often provide ways of filtering results. Aside from entering special sub-domains (images, shopping, etc.), Google allows searchers to filter results to groups such as: recent results, results with pictures in them, previously visited results, new results, and many more. These types of filters are single web links that can be selected one at a time. It is also becoming increasingly popular, with developments in Web languages, to have dynamic filters like sliders and checkboxes (for multiple selections). Originally studied in the early 1990s [2], as demonstrated by the FilmFinder interface shown in Figure 4.13, dynamic query filters that immediately affect the display of results have been shown to help searchers express their needs more quickly and effectively. We now see examples of this on many modern systems, with an increasing presence on websites such as Globrix[16] and Volkswagen[17] (Figure 4.14). Thinking more about the influence of technology and metadata, dynamic filters like sliders are more effective when data is continuous (like price, height, weight, etc.) rather than when data is in categories.

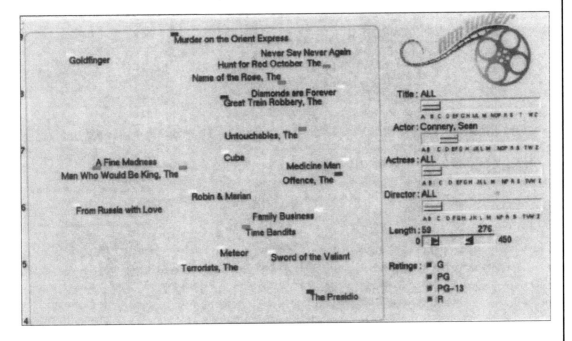

Figure 4.13: Dynamic query filters were shown to help searchers fine information faster; taken from [1].

It is not only sliders that can be used to have an immediate impact on results. When searching through music in iTunes, for example, the collection and browsing columns are immediately filtered

[16]http://www.globrix.com/
[17]http://www.volkswagen.co.uk/#/new/

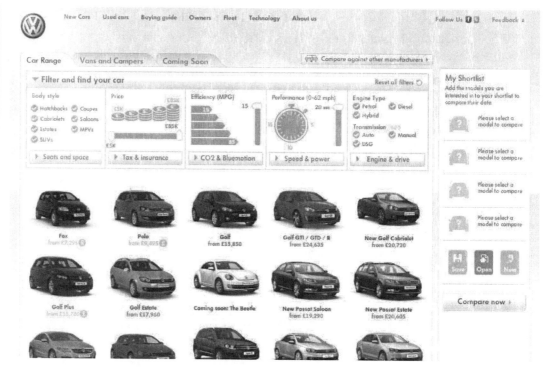

Figure 4.14: Volkswagen use a dynamic and interactive set of sliders and checkboxes to filter their cars.

by the search terms. This reduces the space through which searchers have to browse, where research has indicated that searchers often feel happier making decisions within a smaller set of options [144].

4.2.5   GROUPING

Another approach to re-ordering results is to group the results by their type or by one facet of metadata. The Flamenco browser [217], however, allows searchers to interactively group results by any of the facets of metadata available in the system. Similarly, the digital library provided by the Association for Computing Machinery[18] (ACM) allows searchers to group results by publication venue, year, or institution. These two examples provide the searchers with clear *Control* over the results, but to maintain a sense of consistency (one of Nielsen's heuristics) systems must provide default groupings. The iTunes store, as shown in Figure 4.15 for example, organises results by whether they are matched by Artist, Album, or Song Title, etc. Searchers are then able to focus in on the results that more closely match their intentions.

---

[18]http://portal.acm.org

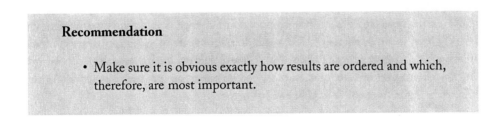

Figure 4.15: The iTunes store groups results by their different types, here showing Artists, Songs, and Podcasts matching 'The Sways.'

The integration of different types of results has also been applied to web search, often called Aggregated Search, where different types of results are grouped within one page. While researchers have been studying the most effective way of integrating different types of results in one screen [185], large web search engines typically integrate different types into the single ordered list so that it is clear which are the most relevant. To investigate further, the Yahoo! Alpha[19] project was experimenting with providing aggregated resources in a variety of different forms. Notably, however, new start-ups like Cuil[20] originally displayed results in a grid, but recently returned to a single ordered list of results so that it was clear which was the most relevant result.

> **Recommendation**
>
> • Make sure it is obvious exactly how results are ordered and which, therefore, are most important.

---

[19]http://au.alpha.yahoo.com/
[20]http://cuil.com–was started by ex-Google employees

These default and passive forms of groupings begin to branch into how results are presented and so can also be considered as an *Informational* feature. Similarly, so do table-based sorting options. We continue in the next section to discuss the wide range of features that are primarily *Informational*.

4.3   INFORMATIONAL FEATURES

Informational features typically relate to how results are organised and displayed. The sections above, however, have made note that *Control* features that present metadata in SUIs can also be *Informational*. Facets, for example, can help inform the searcher about the types of data and results that are available within a system. This section focuses on features that are primarily *Informational*.

4.3.1   STANDARD RESULTS LISTS

Although they can appear fairly simple, the way we present individual results in a Search Engine Results Page (SERP) has been very well researched, but not all of the ideas have stuck. Notably, for example, SUIs rarely include explicit ranking numbers that indicate how relevant they are to a search. Returning to Nielsen's heuristics, such numbers do not represent familiar language to searchers, as they represent the score of an algorithm that most searchers will not understand. A simple ordered list form top-to-bottom, however, provides a clear ordering that searchers understand intuitively.

> **Recommendation**
>
> - Avoid unnecessary information, which can be distracting during search.

Web search results, as shown in Figure 4.16 typically include: (1) the title of the result, (2) a snippet of text from the result, and (3) the URL for the result. Their size, content, and structure, however, have been subject to subtle changes over the years. In 2011, for example, Google moved the URL of the result to be above the snippet. This change was perhaps motivated by research that has highlighted the importance of URLs and domain names in making trust judgements [116, 171]. Further in 2011, researchers at Yahoo! also showed that enhancing search results with features such as multimedia and actionable links, both discussed below, are preferred by searchers and can lead to significantly higher rates of being clicked [68].

Clearly, the representation of different types of media is different. Image Search results typically do not have a snippet of text, but some do have information about the size and filetype. Similarly, in retail domains, results may have pricing information and ratings. The rest of this section discusses influences on these designs.

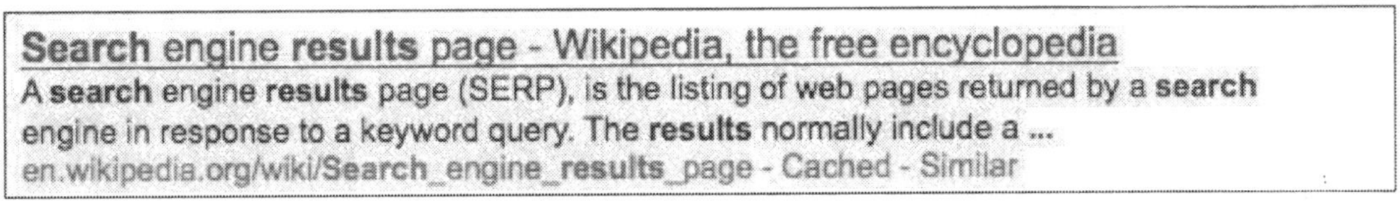

> **Search** engine **results** page - Wikipedia, the free encyclopedia
> A **search** engine **results** page (SERP), is the listing of web pages returned by a **search** engine in response to a keyword query. The **results** normally include a ...
> en.wikipedia.org/wiki/Search_engine_results_page - Cached - Similar

Figure 4.16: A typical example of a single result in a Google results page from 2010.

Snippets

The design of text snippets alone has been carefully studied. Informally, two lines are typically chosen as the optimal balance of communicating useful information and including as many results as possible above the first-scroll point. More formal research [204] has shown that these snippets are best designed if they contain the sentences that include the terms that were searched for, usually highlighting in some way the search terms. This example can be seen throughout the web search figures included in this book.

> **Recommendation**
>
> • Searchers rarely scroll, so get 'important' information above the first-scroll point.

Variations in the size of these snippets are rare. Research by Paek et al. investigated whether longer snippets should be made available on demand as the searcher hovers the mouse over a result [146]. Paek et al. discovered that, although the extended snippets were popular, the UX of hovering and waiting reduced the utility of the SUI feature; searchers were rarely seen to hover and expand the snippets. We do see, however, examples on websites, like Bing's Ciao[21] (Figure 4.17), where the end of a short snippet is followed by a 'more' button; this provides a fast and optional extension to individual snippets, similar to the recommendations of Paek et al. [146]. As a more global layout option, Google also allows searchers to change their search settings to see extended snippets. The choice, however, applies to all or none of the results.

Usable Information

One good timesaving principle is to let searchers actively use results, within the SERP, as and when they find them. There is no reason, for example, that shoppers should have visit results individually in order to add them to a cart, especially if they already know what they want. Sainsbury's[22], for example,

---

[21]http://www.ciao.co.uk/
[22]http://www.sainsburys.co.uk/

Figure 4.17: Snippets in Ciao's search results can be extended using the 'more' link.

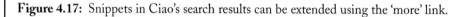

Figure 4.18: Results in Sainsbury's groceries search can be added to the shopping basket without having to leave the search page.

allows searchers to add items to their cart from the SERP, as shown in Figure 4.18. If searchers are unsure if an item is right for them, however, they can view a page with more information about each product, and buy from there too. Ciao!, in Figure 4.17, also has a range of usable links in their results, including links directly to reviews, pricing options, and to the category that an item belongs in. In Google Image Search, there is a usable link that turns any result into a new search for 'Similar Images,' as discussed in the Query-by-example section above. Further, searchers may now '+1' a result in a Google SERP, without affecting or interrupting their search. Finally, searching in Spotify[23] provides a number of usable links in their search results. While viewing a list of tracks that match a search, as in Figure 4.19, searchers can: use the star to favourite a track, buy the track, and

[23]http://www.spotify.com/

visit pages about the artist and the album, rather than having to go to a page about that album or song directly.

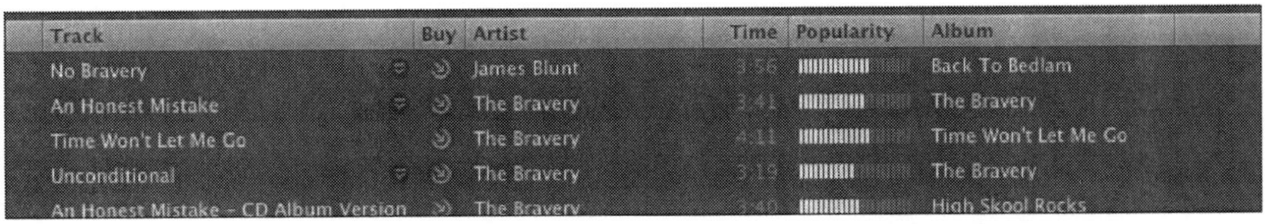

Track	Buy	Artist	Time	Popularity	Album
No Bravery		James Blunt	3:56		Back To Bedlam
An Honest Mistake		The Bravery	3:41		The Bravery
Time Won't Let Me Go		The Bravery	4:11		Time Won't Let Me Go
Unconditional		The Bravery	3:19		The Bravery
An Honest Mistake – CD Album Version		The Bravery	3:40		High Skool Rocks

Figure 4.19: Spotify users can save, buy, and access specific sub-pages when searching for music.

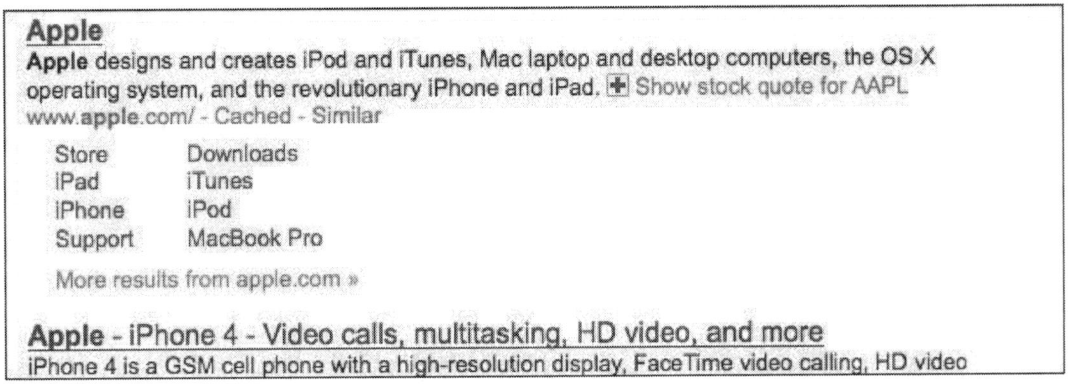

Figure 4.20: Google helps searchers jump directly to certain key pages within more popular websites.

Another type of timesaving usable information, provided for more popular website is to provide 'Deep Links,' as shown in Figure 4.20. Deep links allow searchers to jump directly to certain parts of results. In Figure 4.20, searchers can jump to the store, technical support, or pages about significant products within the Apple website. By allowing searchers to take shortcuts, deep links support searchers in Nielsen's Flexibility and Efficiency heuristic. In Figure 4.17, Ciao also provides deep links for searchers to jump straight to the reviews, for example, within the page about an individual result.

Images

While research suggests that people can make snap-judgements about the quality and professionalism of websites based on how they look [221], research indicates that thumbnails often only provide valuable support when searchers can recognise websites (e.g., [190]). Exceptions to this rule are when visual features are important, such as searching for pictures and choosing between purchasable products. Consequently, thumbnails are typically recommended for when searchers may be re-finding

Recommendation

- Provide actionable features in the SERP results directly so that searchers do not have to interrupt their search.

previously seen websites. When thumbnails are used, research recommends that thumbnails be approximately 200 pixels wide to optimise for recognition [90]. Recognising pages, especially within one website (e.g., Wikipedia) where all results may look the same however, can be difficult. Consequently, several approaches have been taken to make thumbnails more expressive or informative. Woodruff et al. compared several variations on thumbnails, including enhanced thumbnails (shown in Figure 4.21 (a) and (b)) where key terms are expanded within the thumbnail, which in their study reduced total search time and performed the most consistently across different types of pages types (e.g., textual, image, news, product pages) [214, 215].

Taking the concept of turning thumbnails into visual versions of text snippets, Teevan et al. created Visual Snippets [190], which took the colour theme, key image, and key text from a website to create an abstraction of the page, shown in Figure 4.21 (c) and (d). While normal text snippets were shown to be best for first-time search, and normal thumbnails for re-finding, Visual Snippets were shown to support searchers best across both conditions. Further, they noted that Visual Snippets could be smaller than the recommended size of thumbnails, matching more closely the size of a normal result on a SERP. Given their propensity to be quickly recognised, Morgan and Wilson experimented with putting a rack of thumbnails at the top of each SERP, allowing participants to quickly identify pages they had seen before [127]. Morgan and Wilson concluded, however, that the separation of thumbnails from their associated textual results was more frequently disruptive than helpful, creating a poor overall UX for the searcher.

In practice, the inclusion of images or thumbnails in or around search results has been unstable for a long time. While they are critical, of course, for image search, and useful in product or book searches (e.g., Figure 4.14, Figure 4.17, and Figure 4.18), they have been tried and tested in many forms within web search engines. Aside from experimenting with thumbnails on the left and right, Ask Jeeves innovated in 2004 by allowing searchers to show individual thumbnails on-demand using a small binocular icon next to search results. A report by VeriTest studied 87 participants and concluded that searchers performed better in some tasks, examined more results, and showed subjective preference for the version with binoculars [28]. While Ask Jeeves has since dropped this functionality, Google has introduced a similar magnified glass icon, which can be used to display images that are very similar to Woodruff's Enhanced Thumbnails [215], as shown in Figure 4.22.

(a) Enhanced Thumbnail   (b) Enhanced Text Thumbnail

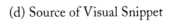

(c) Visual Snippet   (d) Source of Visual Snippet

Figure 4.21: Examples of Web search thumbnails that have been augmented; images from [215] and [190].

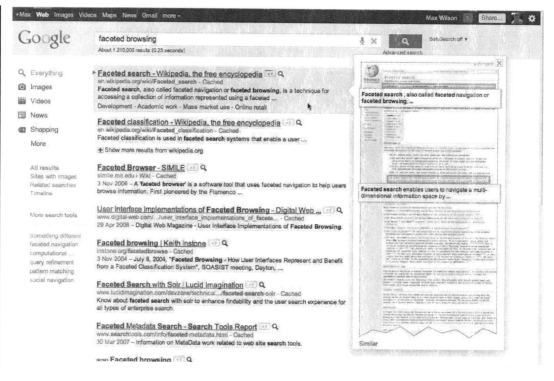

Figure 4.22: Enhanced page thumbnails, similar to those proposed by Woodruff et al. [215], can be displayed by clicking on the magnified glass icon by Google search results.

Previews

Previews allow searchers to see additional information when making a decision about a search result. Like the inclusion of deep links, or usable information within results, previews allow searchers to be more efficient and avoid viewing results unnecessarily. There are many types of previews that can exist in search results: additional text, additional images, other media, etc. In Twitter and Facebook (see Figure 4.23) searchers can hover over usernames and see some primary information about them. One of the more interactive advances that Bing has made over Google is to provide informational previews for each result, which can be activated by hovering over the arrow on the right-hand side. Bing uses this preview method to provide extended snippets and deep links without deviating too far from the standard presentation of results.

The examples so far provide extra information about individual results. Research has also investigated the provision of example previews for higher-level categories. In the mSpace faceted browser mentioned above, searchers can play example audio files that are in some way representative of the category [168]. Hearing an example of Baroque classical music, for example, may help searchers

(a) Hovering over the author of a tweet causes a preview pop-up with more information about them.

(b) I have 10 mutual friends with the user; used with permission.

Figure 4.23: Examples of Previews often provided when hovering over links.

better understand that type of music. While we do not see these category-level previews so frequently, enterprise search vendors, such as Endeca [101], are investigating these techniques to give people summative textual overviews of companies or product categories.

Recommendation

- Images and Previews can help searchers make better browsing decisions.

Relevance Information

As briefly mentioned above, although the relevance of a document is often calculated in a very detailed manner, we rarely see the presentation of a 'relevance score' by results. Everyday searchers will not know what they mean, or how they are calculated, and so relevance scores provide very little benefit over the implicit ranking of the results list. We can, however, communicate relevance in other ways. Clearly, when sorting by specific criteria, such as average rating or number of downloads, we can clearly demonstrate these meaningful values in the order that we display results. The most common form of conveying 'relevance' itself, however, is to abstract the information in some way. We frequently see examples of bar chart style indicators to communicate the relevance of a result. Figure 4.24 shows examples of bar chart style relevance indicators found with Apple's email software and PDF reader. Such relevance judgement can be more specifically helpful when results are sorted by a different dimension, as in Figure 4.24 (a). Abstracting in this form turns relevance information into a language that is clear for the searcher, matching the heuristic principle about avoiding jargon and technical terms.

One early example of abstracted relevance, called TileBars [78] (Figure 4.25), tried to provide detailed information about the relevance of documents, and the parts within the document. Each row of a TileBar represents the elements of the query (3 terms in this case). The length of a TileBar, instead of representing an overall relevance like a bar chart, represents the length of a document. Each horizontally aligned square represents a section of the document. The colour intensity of each square indicates the relevance of that section of the document to that term in the query. Consequently, searchers get a sense of both the size of the document, and the amount of it that is relevant to the different elements of the search criteria. Although we do not see TileBars, as originally designed, frequently in modern SUIs, we see examples of the TileBar principles in many systems. The experimental HotMap [81] interface, shown in Figure 4.26, provides a rotated TileBar-style view over the results of a search. Each red square indicates the relevance of a result to one individual term in the keyword search. Other systems, including the early prototype by Veerasamy and Belkin [198], shown in Figure 4.27, used vertical bars to show how related different results were to different search terms.

The InfoCrystal system [181] (Figure 4.28) provided a unique visualisation for conveying relevance of results to different search terms. InfoCrystal creates a shape with as many corners as search terms. Additional smaller shapes with fewer corners, placed within that shape, indicate how many results are relevant to combinations of these terms, where the number of sides and their colours indicate which search terms the documents are relevant to. Using this system, which can be hierarchically nested, searchers can determine the set of results that are relevant to different combinations of search terms. The next section describes many more examples of how 2D spaces have been used to display and visualise results.

4.3.2 2D DISPLAYS OF RESULTS

Moving into the realms of Information Visualisation, many search systems have explored the use of both horizontal and vertical dimensions to present results. Most commonly, we see search results displayed as a grid (e.g., Image search), where we expect (at least in the western world) relevance to go left-to-right across rows, and repeating for all subsequent rows. This, however, does not so much use two dimensions to present results, as present one dimension of relevance in a different layout. Below we look at loosely organised and structured use of 2-Dimensions to display results.

Loosely Organised 2D Spaces

Embodying the idea of clustering, many 2D displays have been used to present results spatially in how relevant they are each other. 'Self-Organising Maps' (SOMs [100]), for example, cluster results around key terms, as shown in Figure 4.29. The layout is determined by the content of the

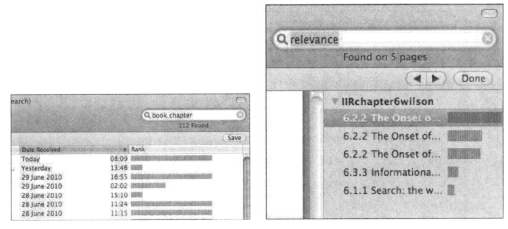

(a) Searching for email Apple Mail.

(b) Searching for terms in a PDF using Apple Preview.

Figure 4.24: Examples of indicating relevance with a bar chart in Apple OS X.

Figure 4.25: TileBars indicate how relevant different portions of a result (columns) are to each search constraint (rows); image from [78].

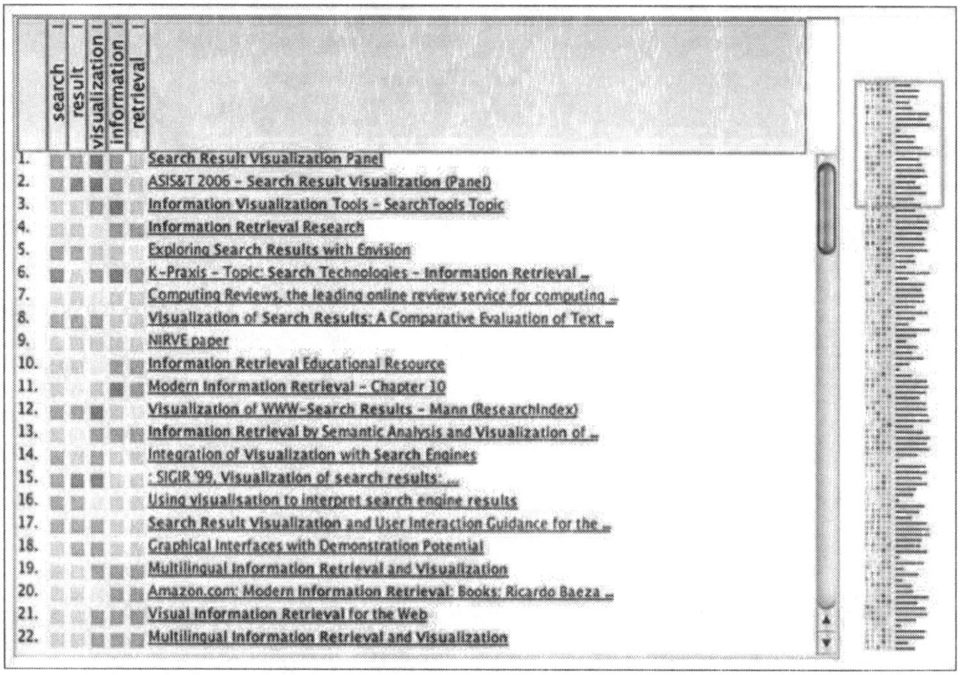

Figure 4.26: HotMap indicates how relevant each result is to the search terms in use.

documents, and the positioning tries to find the optimal distance from the relevance scores between documents. SOMs are based on neural network algorithms, and they can create a layout that can look like a surface with mounds where there are many similar results. Colour intensity, like with TileBars, is often used to create this effect. SOMs can visualise large numbers of results, where the WebSOM in Figure 4.29 is showing one million documents. Searchers can zoom in on areas of the map and select individual documents that are represented as nodes.

Similarly, there are several variations of cluster-based visualisations, where the emphasis in the visualisation is on the connections between clusters, rather than the clusters themselves. The ClusterMap[24] visualisation, shown in Figure 4.30, visualises the cluster connections by highlighting both the topics in the documents, and documents that share multiple topics. The graph shows results that are related to four categories, where 24 documents, for example, are about both Microsoft Word Documents and RDF[25].

There are many alternatives to these examples, including the Hyperbolic Browser [111], which was observed to make the best use of space to display hierarchically structured data. One limitation that these loosely organised visualisations all share is that it can be hard to know where results are.

---

[24]http://www.aduna-software.com/

[25]RDF is a data markup language for the Semantic Web that adds relationships to XML.

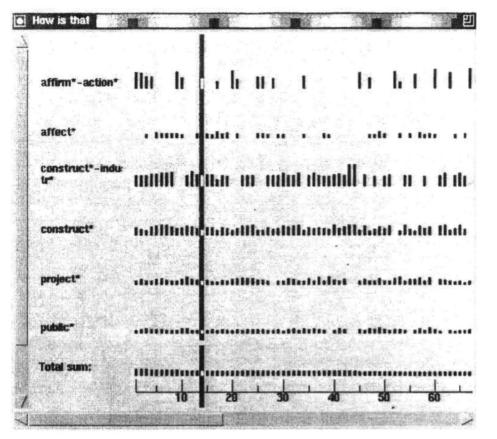

Figure 4.27: Veerasamy and Shneiderman visualized search results (shown as 1-70) along the bottom row, by how they related to different search terms. Image from [198].

They are ordered by how they related to certain topics and each other, and this order may not be clear to the person searching. Further, the order in which results are encountered by some of the layout algorithms, for example, can mean the same result is placed in a different part of the space. Similarly, research into the layout of tag clouds showed that predictable orders, such as alphabetical order, were important for searchers [170]. Ultimately, if a searcher has an idea of what they are looking for, it can take longer to visually search for it in large topically organised layouts, than if it were to appear in a logical order. One alternative is to let searchers control and manipulate these 2D spaces.

Searcher Organised 2D Spaces
In order to make 2D spaces more meaningful for searchers, research has investigated search methods that are similar to the way we can view folders in both Windows and the Mac OS X operating

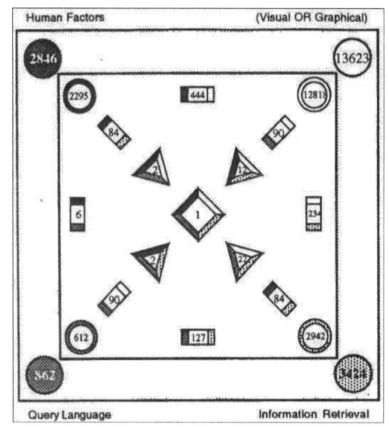

Figure 4.28: InfoCrystal uses concentric shapes to highlight how many documents are relevant to different combinations of search terms. Image from [181]. Figure Copyright © 1993 IEEE. Used with permission.

systems. Early research into the TopicShop SUI [3], shown in Figure 4.31, found that searchers were better able re-find useful websites when they were able to manipulate documents in a 2D space. Participants were able to use spatial memory to remember where they had left documents. Giving searchers control over such spaces allows them to make them personally meaningful. More recently examples, such as the CombinFormation system [97], allow searchers to collect images and text from the web and organise them spatially. These layout methods, however, still depend on recall, rather than recognition, thus conflicting one of the key design UI principles.

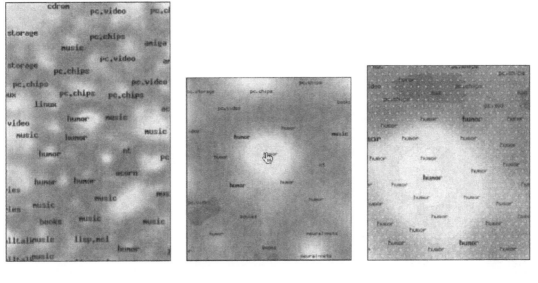

| (a) Overview Level | (b) First Zoom | (c) Second Zoom |

Figure 4.29: The WebSOM organizes newsgroups by how they relate to each other.

Structured 2D Spaces

When results are classified by categories and facets, these metadata can be used to give a 2D space a clear structure that searchers can *recognise*. Further, clear meaningful axes is a standard design recommendation for any visualisation [195]. The TreeMap visualisation [174], shown in Figure 4.32, shows results in based in the hierarchy of categorisation. The 2D space is divided among the top-level categories, in proportion to the volume of results they include. Each of these top-level groups is, in turn, divided among its sub-categories, where the size of those is determined by the number of results they include. Where the space is first divided into columns, each column is then divided into rows. The process continues recursively. This process provides a top-down view of a hierarchy, and the results within them. Colour can often be used to highlight results within this layout, according to a different dimension, such as price. Although circular versions of TreeMaps have been explored, they are generally considered to be less efficient and less clear about volume than the original design[26].

While TreeMaps provide a more structured layout, it can still be hard to visually search through the results. To provide an even clearer view of a 2D space, some visualisations allocate specific facets of metadata to each of the two dimensions. If one dimension shows price, and another shows quality, then searchers can easily identify the cheapest results that match a certain quality. Similarly, searchers could easily identify the most highly rated camera in a given price bracket. Several systems, including the Envision system [141], the List and Matrix browsers [108], and the GRIDL SUI [175] shown

[26]http://lip.sourceforge.net/ctreemap.html

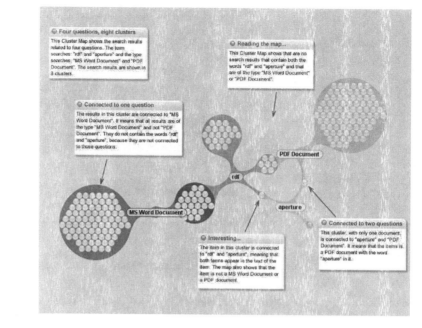

Figure 4.30: ClusterMap shows how documents relate to one or more categories.

in Figure 4.33, use both the vertical and horizontal dimensions for specific facets of metadata, and display results in the marked-out space that match. Each dimension is browseable and/or filterable, so that searchers can explore the data with different facets, or at different granularities. Often, the visualisation within each square can be changed.

Timelines are an example of a structured 2D space, where one dimensions is clearly attributed to time. Often the second axis is given another meaning, whether it is a categorisation or a measurement scale. The optional timeline that can be shown in Google shows number of results in the vertical axis, and divides the horizontal axis into portions of the specified time period. Timelines and other visualisations, however, can often struggle to show detail while providing a clear overview. The Simile Timeline[27], since adopted by Google Visualisations[28], support focus and context by stacking different granularity views of the same timeline on top of each other. An early prototype called the Perspective Wall [118], achieved this by using the third dimension to have panels on the left and right of the timeline that show time stretching out into the distance. Focusing more on the amount of information that could be conveyed in 2D, André et al. investigated representations of interrelated faceted metadata within their Continuum timeline [6], shown in Figure 4.34. Similarly, the LifeLines timeline visualisation focused on optimising timelines to show relationships [152] and

---

[27]http://www.simile-widgets.org/timeline
[28]http://code.google.com/apis/chart/

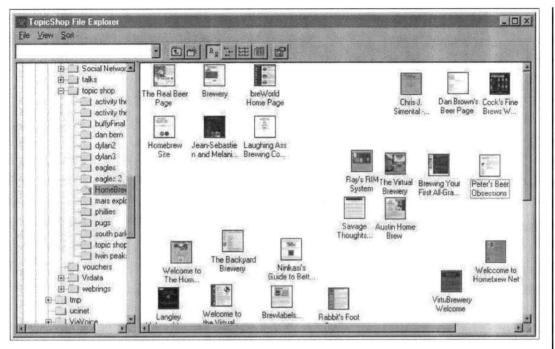

Figure 4.31: TopicShop allows users to organise web search results in a 2D space; image from [3].

compare temporal patterns [213]. Maps are another common kind of structured 2D space, where not only do the 2 dimensions represent longitude and latitude, but also the land-masses are extremely familiar to most users.

4.3.3   3D DISPLAYS OF RESULTS

Naturally, in line with the progression in technology, many visualisations have been extended into 3D versions in order to increase the number of results that can be displayed. 3D perspectives have been added to SOMs, and the DataMountain prototype [159] (Figure 4.35) adds a third dimension to the user-controlled spatial layouts. The BumpTop software[29], now acquired by Google, provided a 3D layout and mock-physics to allow users to organise files on the desktop. Consequently, the BumpTop software provided a 3D space very similar to the DataMountain and TopicShop SUIs. Further, the exploration of a hierarchical data was extended in a 3D space, in a design called Cone Trees [160], as shown in Figure 4.36. Cone Trees were later used in the Cat-a-Cone prototype to enrich the 3D display with multiple simultaneous categorisations, where any found results would be highlighted in multiple places across the full context of the hierarchy [79]. The Hyperbolic Tree

---

[29]http://www.bumptop.com

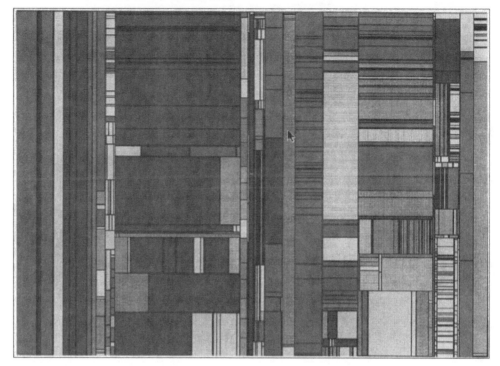

Figure 4.32: This TreeMap shows 850 files in a 4 level hierarchy, where colours represent different file types. Each level of the hierarchy is broken down by changing divisions between columns and rows; image from [174].

Recommendation

- Make sure the dimensions and layout of a visualisation are clear and intuitive to the searcher.

was also expanded to use three dimensions [134]. There have also been examples of stars-in-space style visualisations, which searchers can explore in 3D.

While many of these ideas are exciting, they have rarely been shown to provide significant improvements to searchers (e.g., [38, 172]). Essentially, the overhead of manipulating and navigating through a 3D space currently overrides the benefits of adding the third Dimension. 3D visualisations also typically involve overlap and occlusion. Further, some research has shown that around 25% of

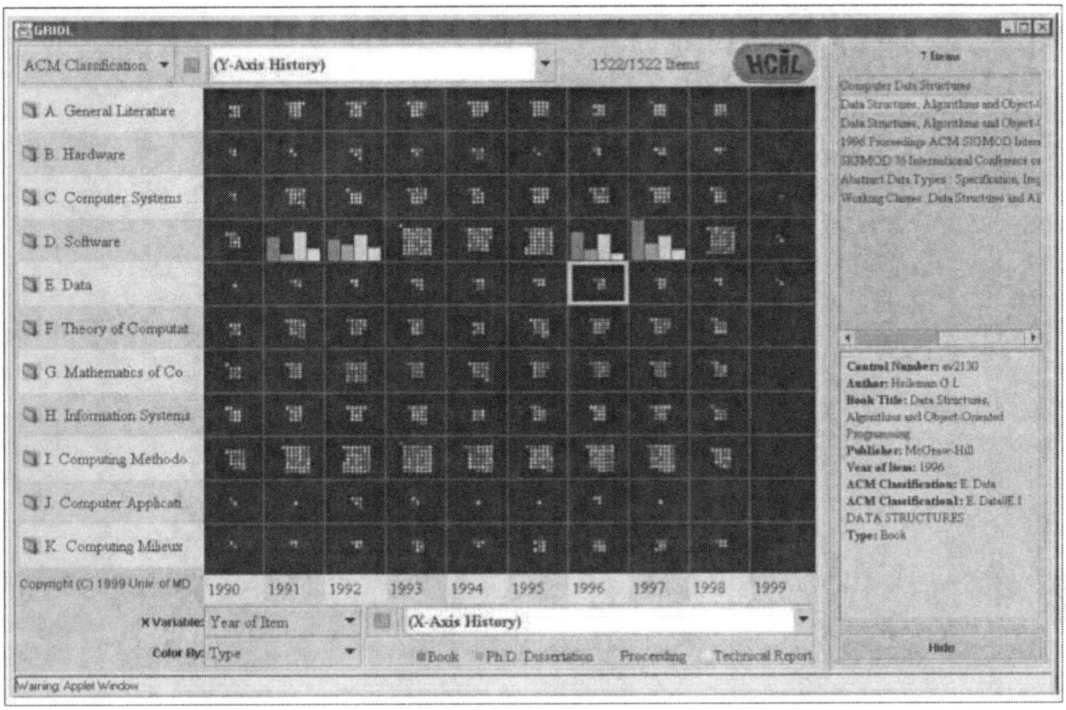

Figure 4.33: GRIDL visualises information by creating a grid between two clearly identifiable, but changeable category dimensions; image from [175].

the population find it hard to perceive 3D graphics on a 2D display [126]. Finally, the technology to deliver 3D environments on the web is still somewhat limited. Research, however, continues into 3D environments and 3D controls, and so it is not beyond the realm of possibility that we'll see more common use of 3D result spaces in future systems, especially with the onset of multi-touch devices and 3D displays.

4.3.4   ADDITIONAL INFORMATIONAL FEATURES

So far this section on *Informational* features has mainly discussed the way in which we can display results. There are many other secondary *Informational* features that guide and support searchers in a SUI. These details are often very subtle and often determined by larger UX concerns. Placed correctly, however, they can provide important information at just the right time, often to improve the overall UX of a SUI.

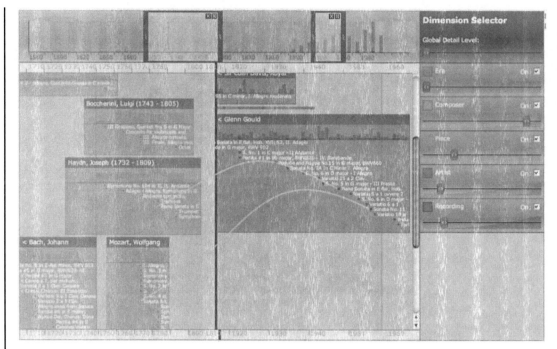

Figure 4.34: Continuum is a timeline that can display hierarchical relations, display temporal relationships, use split views, control which facets of metadata are displayed, and their level of detail using sliders. Image from [6].

Guiding Numbers

Although SUIs typically avoid displaying numeric representations of relevance scores, there are many other times when the simple inclusion of numbers can help guide searchers. Google rather crudely provides numbers relating to how many results were found, and almost brags about the short time in which they were found (see #10 in Figure 1.1). The number of results can provide insight for searchers into the size of the space they are trying to explore, which is emphasised when there are very few pages of results.

The use of guiding numbers can be particular valuable in browsing and faceted SUIs, where they may tell searchers how many related items there are in a category. Searchers may, for example, choose to check a smaller category before trawling through the results of a larger category. The use of guiding numbers is fairly common, and shown throughout many of the figures in this book. One example is the number of items relating to an option in a facet, as shown clearly in Figure 4.8. Such indicators have also been represented visually, similar to a bar chart, to guide searchers away from empty result sets [48, 219].

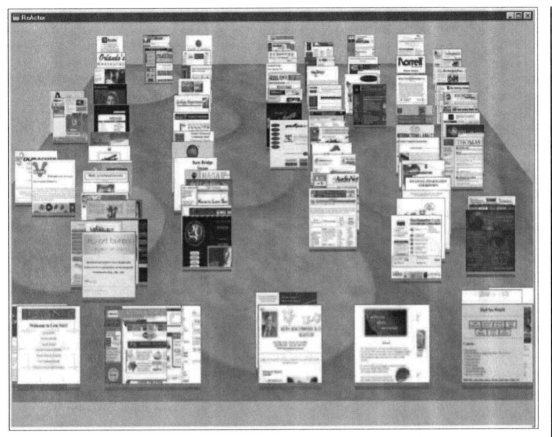

Figure 4.35: Data Mountain provides a 3D space for searchers to manually organise the results; image from [159].

Zero-Click Information

It is becoming increasingly common for search systems to provide zero-click information. Search engines often automatically answer some simple queries. Searchers can get exchange rate information by entering '£ 1 in $,' and most search engines provide the answer above the normal 10 results. Similarly, search engines typically respond to mathematical sums, and convert measurements into different scales. DuckDuckGo[30] focuses on the idea of zero-click information by trying to provide an answer or a piece of information in every SERP. DuckDuckGo may provide, for example, a short biography about a person, or the definition of a word, or the current weather, by identifying entities in the query, such as person, a term, or a location. Such instant answers to queries correspond to

---

[30]http://duckduckgo.com/

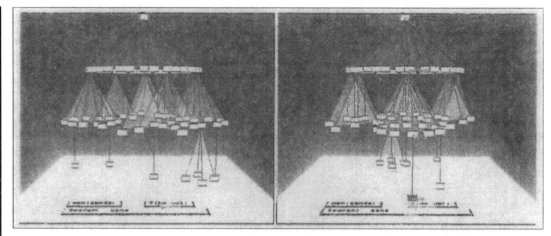

Figure 4.36: Cone Trees use the 3D space to visualise a much broader hierarchy. Image from [160].

> **Recommendation**
>
> • Guiding numbers help searchers to make better browsing decisions.

Nielsen's Efficiency heuristic, but often involve sacrificing screen space that would otherwise be used for more results.

Chilton and Teevan [34] studied Bing's query logs and found that phonebook information and news headlines accounted for 89% of all zero-click answers experienced by searchers, and the ones that were frequently reused by searchers were phonebook information, flight status, news headlines, and weather details. For certain types of answer, searchers would rarely then interact with the search results, including: sport updates, flight details, and weather details. Li et al. refer to this lack of subsequent interaction as 'Good Abandonment' during search [114], where it is positive that a searcher did not click on any results.

Signposting

In line with Nielsen's *Visibility* heuristic, it is important to make sure searchers 'know where they are.' The introduction above already mentioned maintaining a query in a search box above the results, so that searchers know exactly what is being searched for. Similarly, the section on faceted browsing mentioned examples where the choices searchers have made are still clearly visible, rather than removed to allocate more space to new decisions. One common form of signposting for more browsing-oriented SUIs is to provide a breadcrumb trail, as shown in Figure 4.37. Breadcrumbs show where in the hierarchy, for example, searchers are, and each level of the hierarchy shown in Figure 4.37 (a) is also a hyperlink to go back to that level. Consequently, searchers can easily jump up to previous levels; again providing a shortcut for Nielsen's *Control and Freedom* and *Efficiency* heuristics. Faceted browsers that do hide previous decisions often provide a breadcrumb of previous choices that are ordered by decision made, rather than imposed by a hierarchy. For faceted SUIs, items in the breadcrumbs are constraints that can be removed, as in Figure 4.37 (b), rather than to allow searchers to go back to a certain stage of browsing.

(a) Hierarchical breadcrumbs let searchers jump to a high level.

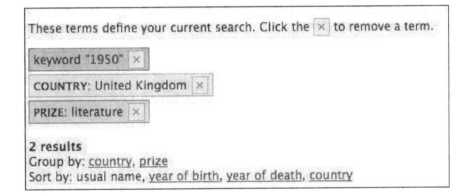

(b) Faceted breadcrumbs let searchers remove individual constraints to broaden their search.

Figure 4.37: Breadcrumbs provide signposts to keep searchers informed about the factors involved in their search.

Pagination

Although subtle, one of the key *Informational* indicators for both the number of results searchers are viewing, and where in the results they are currently browsing, is the pagination control. Pagination has several advantages: it limits load time, it provides natural break/decision points in the browsing process, it provides feedback on how far through the results searchers have reached, and how far they can go. Implementations, however, vary. Most implementations provide a limited series of page links (i.e., pages 1–10), preceded by a 'Previous' and followed by a 'Next,' where the 10 linked pages shift as the user searches (i.e., pages 5–15). Where the number of results is more exactly know, pagination often also provides 'First' and 'Last' page links, so that searchers can skip back and forth quickly. Simpler implementations often simply list all page numbers over several lines, which can create a poor UX.

More recently, the advances in JavaScript-enhanced web applications have produced examples where the pagination is seamless. AJAX is sometimes used to load the next 10 results, for example, as the searcher nears the end of the current set, and then appends them without re-loading the page. The searcher can then simply continue to scroll rather than move to a different page. Google applies this technique particularly to their image search. Similarly, Twitter and Yoople![31], a socially enhanced web search engine, provide same-page result extension rather than direct pagination.

While a nice example of flexible JavaScript-enhanced SUIs, this type of pagination can lose some of the guidance and feedback information, affecting the overall UX. It can be harder, for example, to return to a result somewhere in the middle of the longer list, than it is to return to, for example, page two. Yoople!, as shown in Figure 4.38, tries to mitigate this loss of feedback by providing bounding boxes for each appended 'page' and links on the right of the screen to jump to certain pages in the long list.

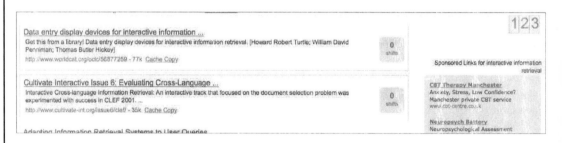

Figure 4.38: Yoople! appends new pages of search results to the same list, marking a page change with bounding boxes.

Animation

The use of animation in any user interface, and thus including SUIs, has had mixed responses over time. Typically, animation requires more advanced web UI development, and can impose longer loading times and thus slower responses. Implicitly, we also often associate animation with annoying pop-ups and adverts. When used appropriately, however, animation can provide informative and

---

[31]http://www.yoople.net/

helpful feedback to searchers. Most notably, animation has been shown to help convey change [15], as it draws attention to the altered object in the interface. Animation was shown to be particularly valuable in conveying change when pivoting between different hierarchy structures in the 3D Polyarchy (multiple hierarchy) visualisation [158]. This valuable use of animation is becoming more common in web interfaces, including Facebook and Twitter. When hovering over a new notification in Facebook, for example, its colour will highlight and fade to show that it has been addressed. Animation should be used carefully and purposefully in SUI design.

> **Recommendation**
>
> • Animation should be used carefully and purposefully to convey a message, such as change.

Social Information

Finally, one common type *Informational* feature, is one that provides the wisdom of the crowds to the searchers. The most common example we see is tag clouds, which provide annotations that others have made over results. Similarly, systems like Digg[32] provide a current 'Hot' list of interesting websites, based on how many people recommend them. Digg is an example of SUIs that log and then use the results that other searchers find, download, or buy. Such systems take value from the actions of their users and provide this value to other searchers. In the front-page of YouTube[33], for example, we see suggested videos that are either the 'Most popular' or 'Videos being watched now.' Further, in Amazon, we see many pieces of crowd-sourced information. Amazon provides reviews and ratings, and information about other items that people have bought together with the item currently being viewed. Amazon even tells searchers if buyers typically end up buying a different item to the one currently being viewed, as shown in Figure 4.39. For searchers with Google accounts, Google results are also highlighted when people in their social networks have shared them or +1'd them. These are two quick examples of *Informational* SUI features that can be *Personalisable*; the next section discusses these types of features in more detail.

4.4   PERSONALISABLE FEATURES

The final type of SUI features to discuss are *Personalisable,* which tailor the search experience to the searcher, either by their actions or by the actions of other searchers who are personally related by some social network. These features typically impact the content displayed in *Informational* features,

---

[32]http://digg.com/
[33]http://www.youtube.com/

Recommendation

- Track and reuse information about the behaviour of a systems searchers.

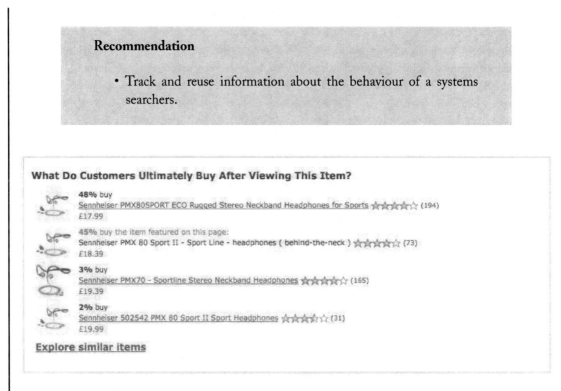

Figure 4.39: Amazon often provides feedback to tell searchers what people typically end up actually buying.

or even the way they are presented. Further, they can affect the *Control* features that are provided. For clarification, there has been a lot of work that has focused on algorithmic personalisation for search, which has a whole book of its own [133]. Instead, this section focuses on different types of *Personalisable* features that appear in a SUI and the impact they can have.

4.4.1   CURRENT-SEARCH PERSONALISATION

The most common type of personalisation found within a single search session, is to provide something like a shopping cart to searchers, or a general collection space. In a recently retired[34] experimental feature, Yahoo! SearchPad (Figure 4.40) provided searchers with a space to collect search results and make notes about them. When activated, SearchPad logged the searches made and the websites visited. When opened, searchers can remove items from the SearchPad, or add notes for themselves or others to read later; SearchPad entries could be saved and emailed.

---

[34]http://help.yahoo.com/l/ph/yahoo/search/searchpad/spad-23.html

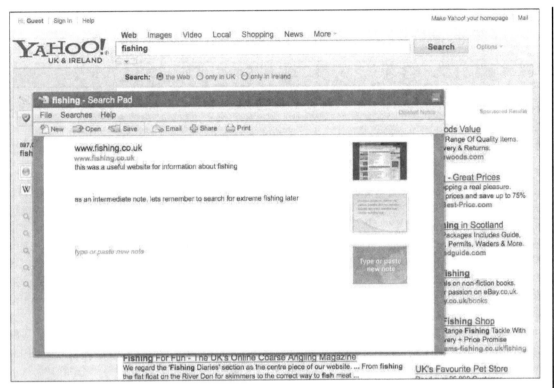

Figure 4.40: Yahoo! SearchPad provided searchers with a space to keep notes and useful results during their search.

Another within-session personalisation feature is to provide a list of recent searches and recent results viewed. Searchers are then quickly able to return to a previous SERP that they thought was useful, or go back to a specific result if they did not manage to find anything better. Figure 4.41 (a) shows an example from PubMed[35], where previous searches are shown, and Figure 4.41 (b) shows a combination of previous searches and viewed results from Amazon. This *Personalisable* feature provides a new type of *Informational* feature. Many other services, including eBay and Bing, provide this feature; many also keep this information for the next time searchers return.

4.4.2 PERSISTENT SEARCH PERSONALISATION

When a SUI can track a searcher over multiple visits, search systems can help searchers continue from where they left off. Multi-session personalisation features can be as simple as remembering settings, like a preference for including thumbnails in results, or remembering the items in a shopping basket.

[35]http://www.ncbi.nlm.nih.gov/pubmed

Recommendation

- Help searchers to return to previously viewed SERPs and results.

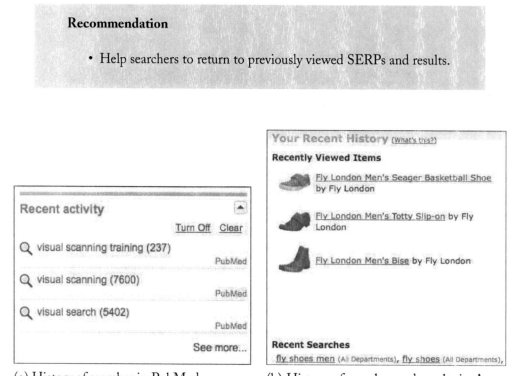

(a) History of searches in PubMed.   (b) History of searches and results in Amazon.

Figure 4.41: SUIs can help searchers get back to previous searches by keeping a history.

Amazon and eBay, for example, assist searchers by recommending items that were viewed in the last session on the home page. Such features can be very helpful in multi-session search scenarios, like planning a holiday or buying a car [95, 117]. When SearchPad was active, it was designed to support such tasks, by helping account holders to resume previous sessions quickly and easily. Similarly, Google extends the idea of a per-session history of queries, by providing their account users with a complete history of their searches and page views (#12 in Figure 1.1). Google also uses this history to modify the *Informational* view of a single search result, by adding the date that a searcher last viewed a result, or indeed how many times they have viewed it (Figure 4.42). Further, Google tells searchers who in their social networks have shared a link or "+1'd" it, as shown in Figure 4.42. The concept of "+1"-ing a website is Google's most recent evolution of highlighting results that searchers like, where previous versions included starring a result, as shown in Figure 4.43, or pressing a button that would also show a certain result at the top of a SERP.

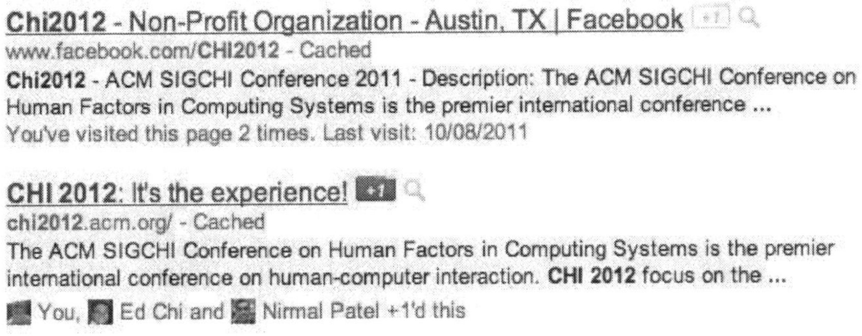

Chi2012 - Non-Profit Organization - Austin, TX | Facebook
www.facebook.com/CHI2012 - Cached
Chi2012 - ACM SIGCHI Conference 2011 - Description: The ACM SIGCHI Conference on
Human Factors in Computing Systems is the premier international conference ...
You've visited this page 2 times. Last visit: 10/08/2011

CHI 2012: It's the experience!
chi2012.acm.org/ - Cached
The ACM SIGCHI Conference on Human Factors in Computing Systems is the premier
international conference on human-computer interaction. CHI 2012 focus on the ...
You, Ed Chi and Nirmal Patel +1'd this

Figure 4.42: Google augments the *Informational* parts of search results with *Personalisable* information about searchers' previous actions and the actions of people in their social networks.

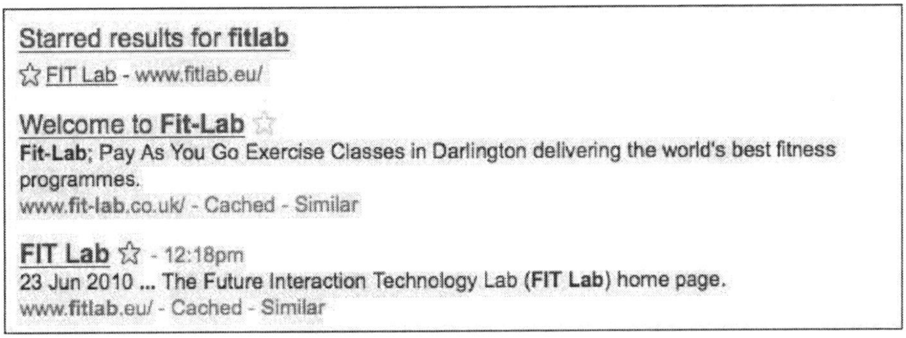

Starred results for fitlab
☆ FIT Lab - www.fitlab.eu/

Welcome to Fit-Lab ☆
Fit-Lab; Pay As You Go Exercise Classes in Darlington delivering the world's best fitness programmes.
www.fit-lab.co.uk/ - Cached - Similar

FIT Lab ☆ - 12:18pm
23 Jun 2010 ... The Future Interaction Technology Lab (FIT Lab) home page.
www.fitlab.eu/ - Cached - Similar

Figure 4.43: For a period preceding early-2011, Google allowed searchers to 'Star' items that they wish to see again. Results are then presented above the main search results.

Many systems provide personalised recommendations, either by previous activity or by the activity of others. Account holders on websites like YouTube and Amazon get personalised information on the front page in place of standard popular results. YouTube makes personalised recommendations based on a searchers prior viewing history. Such features, however, would not be available without appropriate recommender algorithms being implemented in the underlying system, and so represents a good example of how a HCI, UX, and algorithmic work can be interdependent.

4.4.3 SOCIALLY AFFECTED PERSONALISATION

The onset of social networking sites like MySpace, Facebook, Twitter, and now Google+, have provided a new opportunity to leverage more socially aware *Personalisable* features. Research has shown that people particularly trust the opinion of people that they know [64]. More specifically,

> **Recommendation**
>
> - Help searchers to recover their previous search sessions, as they may be back to finish a task.

research indicates that if you know someone, then you are more likely to know if you can trust his or her opinions or choices. Consequently, as per the Figure 4.42, many systems now try to provide information about the searching behaviours of people that a searcher knows. A social-search Question and Answer system called Aardvark[36] links with social networks in order to route questions to people who (a) should know the answer, and (b) already know the searcher. Essentially, Aardvark tries to tell a searcher who, of all the people they know, should be best to answer a question, and then joins them up. Some recent work (e.g., [131, 148]) has shown that many people ask questions of their social network, often for advice, help, or recommendations.

Another service to provide socially *personalisable* is TripAdvisor[37], which allows searchers to find recommended places to stay when travelling. As a first step, TripAdvisor tells a searcher when one of their Facebook friends has been to the location they are currently searching. TripAdvisor then offers searchers the ability to send their friend a message about that place. Further, if a searcher happens to find a hotel that a friend has reviewed, for example, then the reviews are default ordered to 'Friends First,' as shown in Figure 4.44. This is a great example of using *Personalisable* information to improve *Control* examples.

4.5   SUMMARY

This chapter has provided the main review of SUIs and SUI features. Categorised into four types of features (*Input, Control, Informational,* and *Personalisable),* a range of general web search and vertical (within website) search, online search and offline search systems have been discussed. *Input* features typically try to provide either suggestions for keyword searchers, or metadata that can be browsed. *Control* features can affect either the search that was being submitted, or can organise or rearrange the results. A broad range of different *Informational* features have been discussed that present results in linear, 2D, and even some 3D visualisations. *Informational* features, however, can be just as important for supporting awareness and progress monitoring, as well as providing the actual results. Finally, *Personalisable* features were discussed that typically have an impact on the way the prior features behave or affect the information they present. The next chapter covers some a few of the more experimental forms of searching that research is currently focusing on.

---

[36]http://www.vark.com
[37]http://www.tripadvisor.com

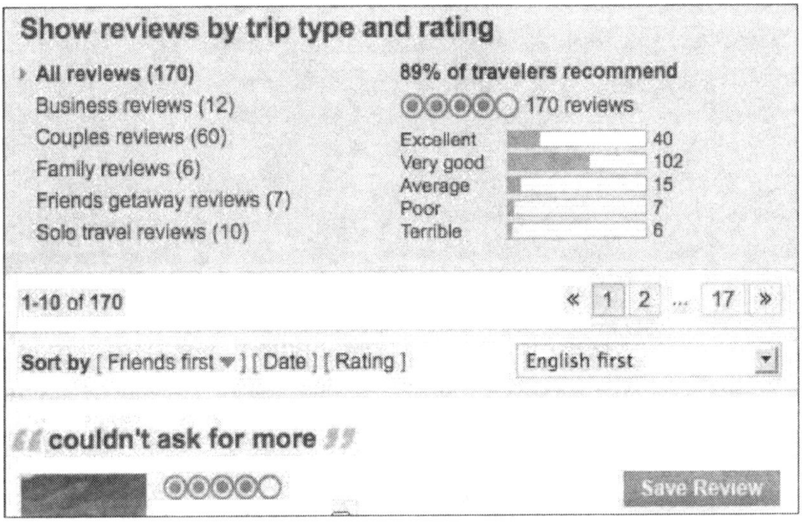

Figure 4.44: TripAdvisor prioritises the reviews from friends in a searcher's social network.

CHAPTER 5

Experimental Search User Interfaces

So far, this book has mostly covered mostly established, deployed, or well-researched SUIs and SUI features, while including examples of some recent advances. There are still, however, many experimental SUIs being created, and novel research areas being studied, such as collaborative searching and use of social media. Google's retired 'Wonder Wheel,' shown in Figure 5.1, is an example of an

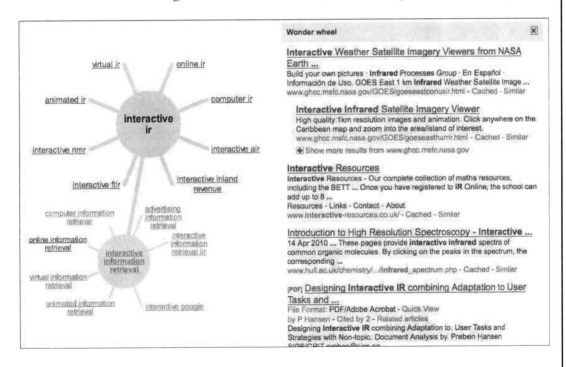

Figure 5.1: Google provides an interactive way of exploring related searches with the experimental Wonder Wheel.

experimental SUI that was deployed but has neither seen widespread use nor published evaluation. The Wonder Wheel presented related queries visually in a circle around the current search terms. Searchers could then click on any of these related queries to change the search. A circle was drawn around the newly chosen query, and the old query-circle was faded slightly. New related queries were placed around the revised search. Searchers could repeat this process, and each selection provided a new set of search results to the right. Essentially, the wonder wheel provided a new visualisation of and an alternative step-wise interaction for interactive query suggestions.

The rest of this section discusses some novel SUI directions, which have produced innovative new designs, but are often yet to appear commercially or to become mature as areas of research.

5.1   COLLABORATIVE SEARCH

Research by Morris, in 2008, identified just how frequently people search collaboratively with one or more friends, suggesting that 80% of people have done so in a pair or group at least once a month [128]. Collaborative searching can involve having someone stand over a searcher's shoulder, or can involve searchers on separate computers. People may collaborate on a search at the same time, or search at different times when convenient to the individuals. Further, people may be searching in the same place, or communicating from a distance. People may be searching collaboratively to plan a group holiday, for example, or be working on a group project. Evans and Chi noted, by synthesising previous research into Information Seeking, that around 30% of searches are prompted by other friends and colleagues, and that around 50% of findings are shared in some form [58]. Golovchinsky et al. noted concurrency, location, explicitness of collaboration, and depth of system support, as four dimensions that model the type of collaboration search being supported [65].

Some recent systems have tried to support collaborative search directly, while other systems have focused on allowing users that are watching to help read results on their phone and suggest new searches [5]. One of the more notable examples, called SearchTogether [129] (Figure 5.2), allows people to collaborate through an extension to Internet Explorer. Searchers can invite friends into their search session. Searchers can see each other's searches and share pages they have found with each other, while together collecting results in a communal report space. Continuing research into collaborative search is looking at, for example, supporting awareness by providing *Informational* features about other peoples search behaviours [173], and providing visualisations to help make sense of what other people have found during their search [149]. Collaborative Information Seeking is a growing area of research, which has been covered in much more detail in a recent book [130].

5.2   REAL-TIME SEARCH AND SOCIAL MEDIA

The rate at which new content is being produced online continues to increase, and has recently promoted a focus on finding the most up-to-date information, over the most relevant. In some circumstances, the most up-to-date information about on-going events is posted by eyewitnesses on social media websites like Twitter, before being reported officially in the news. The idea of real-time search, however, is, at the time of writing this book, still evolving as services like Twitter and Google+ continue to change the landscapes of online communication. After providing a real-time search service for a few years, Google removed it and ended their contract with Twitter. Similarly,

Figure 5.2: SearchTogether allows people to communicate, see each other's searches, and recommend results. (a) integrated messaging, (b) query awareness, (c) current results, (d) recommendation queue, (e) (f) (g) search buttons, (h) page-specific metadata, (i) toolbar, (j) browser. Image from [129].

OneRiot dropped their real-time search engine, shown in Figure 5.3, in 2011. Much is still to be learned about how people search through up-to-date and real-time information from social media sites. In the meantime, however, Teevan et al. noted that common social-media search tasks include: keeping track of events, gaining recommendations, and learning about people [192].

OneRiot's real-time search provided a much more detailed experience compared to Google's scrolling list of recent tweets. Initially, OneRiot scrolled current trends in new content horizontally beneath the search box, preceded by the term 'Breaking.' Upon searching, OneRiot provided results that were recently shared, along with how often they had been shared via social media sites. Although the website results returned aren't necessarily newly published, they were considered 'hot' by how people share and discuss them online. Further, OneRiot provided a sample of the latest tweets about a topic. While OneRiot delivered currently 'hot' content, Google's initial approach, between December 2009 and June 2011, was to include tweets in an auto-updating scroll-box within the

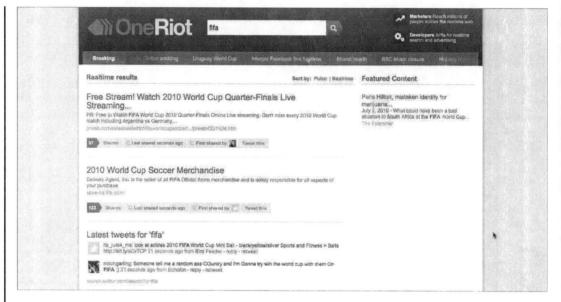

Figure 5.3: OneRiot shows web search results that have been recently shared or discussed by people online.

SERP. Further, Google provided a filter to show the latest results, which used animation to introduce new content into a recency-ordered SERP, as shown in Figure 5.4, such as: tweets, Facebook content, blogs, discussions, and news items.

There are several attempts to understand how best to integrate real-time information into search systems. Eddi, shown in Figure 5.5, is a SUI that is designed to help people manipulate and explore social media updates by time, topic, and source [19]. Others research is still exploring social-media search at an conceptual and algorithmic level, with Hurlock and Wilson identifying 33 characteristics of useful and non-useful tweets [83], and Naveed et al. studying what makes an interesting tweet [136]. Such findings will likely lead to algorithmic improvements before they have an impact on the way real-time and social media SUIs are designed.

5.3   EXPLORATORY SEARCH AND SENSEMAKING

The topic of Exploratory Search is receiving more interest than ever before [202]. Exploratory Search research focuses specifically on SUIs that support the searching scenarios where users are a) unsure of what they are looking for, b) unsure of how to use the current system, or c) unsure about a domain of information they are searching within. Exploratory Search, therefore, takes a view of more challenging tasks. Sensemaking is a process that searches may go through in order to make sense of new information or a large collection of results [44, 45]. Both of these focus heavily

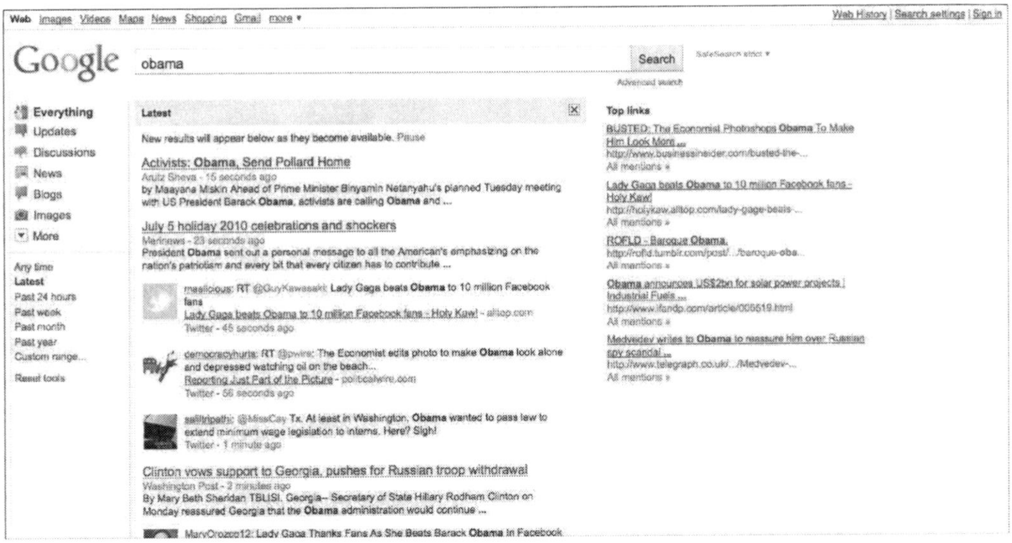

Figure 5.4: Google provides an option to introduce newly published news articles, blogs, discussions, tweets and other social media updates.

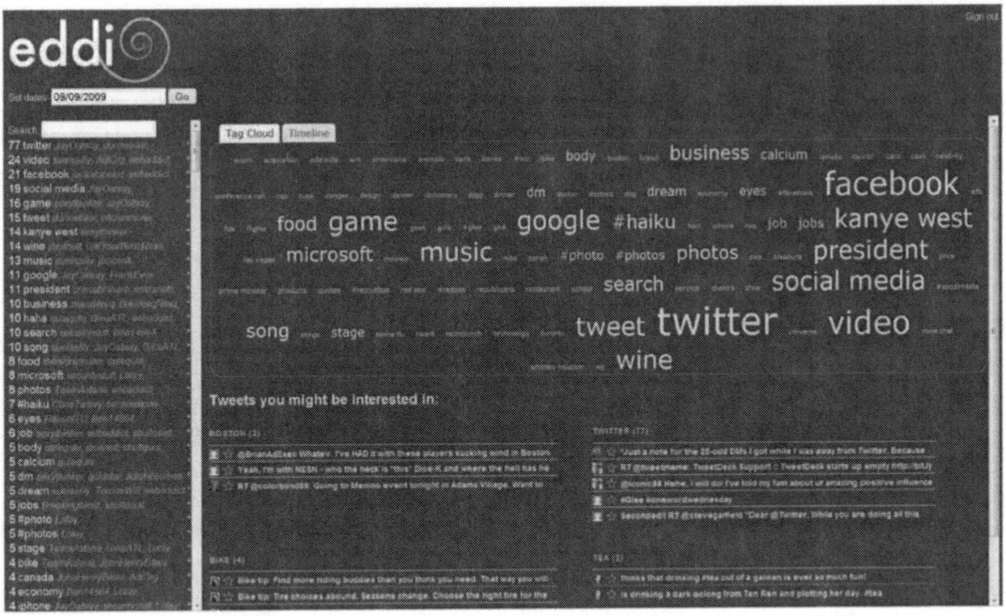

Figure 5.5: Eddi allows searchers to explore real-time social media updates using SUI features like tag-clouds, image from [19].

on human behaviours and so the design of a SUI has a key influence on the types of exploratory behaviours that searches can perform.

Usually, research into Exploratory Search has focused on systems that allow searchers to better explore or browse, however research shows that exploratory behaviour occurs frequently within standard web search engines [203]. The challenge for designing an Exploratory Search, therefore, is providing the right types of SUI features to support the right types of exploratory behaviours. Consequently, a number of the systems described in this book are designed to support more exploratory forms of search. The mSpace browser (Figure 4.7), for example, was designed to help people explore, and the maintained presence of facets is designed to help searchers make sense of the domain of the information. The MrTaggy (Figure 4.9) system was designed to encourage exploration through the use of tag clouds. Another system, called Phlat [40] (Figure 5.6), was designed to be exploratory by providing a set of well-integrated *Control* features that help searchers progressively browse and filter their email and documents.

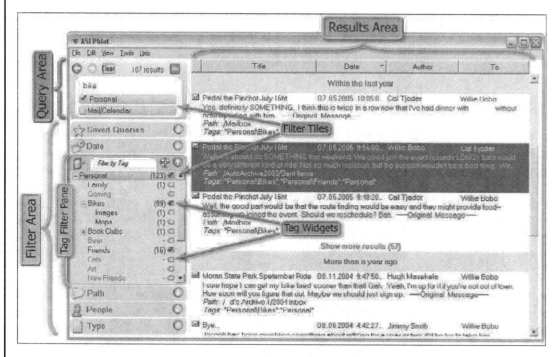

Figure 5.6: Phlat provides a series of filters and tagging to allow users to quickly browse personal collections of documents. Image from [40].

It has been suggested [164] that exploration is well supported by allowing searchers to endlessly follow links (a bit like how we use to 'Surf the Web' before web search engines) from result to result. Systems like Amazon, therefore, support exploratory search and discovery by providing 'similar results' under many different associations: purchases, reviews, searching sessions, company product lists, category, pricing range, recommendations, etc. Each of these types of associations work on the same kind of principle, but each provides a different attribute for searchers to explore the collection. Consequently, it may be considered that the number of different ways that a collection can be explored, rather than the explicit features of the SUI, may be important.

This area of research is relatively open, with many still discussing the nature of scenarios where exploration is likely or preferable, and how to encourage it. Elsweiler et al., for example, have begun to model the real casual-leisure scenarios that people engage in for more pleasurable hedonistic reasons, rather than being information or task driven [57]. Elsweiler et al. highlighted that during these times, people engage in searching behaviours, but intend mainly to spend time, have fun, or discover new and exciting information. These scenarios, such as window-shopping online, browsing Wikipedia, or watching videos on YouTube, may be better evaluated by how engaging they are [142], rather than by how quickly information can be found. More can be read about exploratory search in a book dedicated to the topic [202].

5.4   MOBILE SEARCH

An increasingly important area of search design is for mobile devices, which continue to become more pervasive in our lives. Despite the prominence of mobile technology and the established MobileHCI community [87], the design of mobile SUIs has been under-represented. Further, even though research in 2006 showed notable differences between desktop and mobile search logs [92], follow-up work by Kamvar et al. in 2009 noted that search on high-end mobile devices is becoming increasingly similar to normal web search [93]. Regardless, a diary study in 2008 noted that mobile information needs were weighted towards timely information and trivia, directions, and points of interest [180].

Novel mobile SUIs tend to take advantage of some element of the device's design or the geo-spatiality awareness of the device. FaThumb [94], shown in Figure 5.7, is an example of a Mobile SUI that uses the phone's keypad to help searchers browse through faceted metadata. Each facet was represented by a button between 1 and 9. FaThumb allowed users to perform fast faceted search on a mobile device by making best use of the physical interface. Touch-screen search systems have to optimise for mobile sized screens. mSpace Mobile, for example, implemented a zoomable faceted interface to expose a wider range of metadata to the searcher, allowing the searcher to trade SUI space from one feature to another [205]. Many systems simply fall back on providing a mobile version of their website. Google, however, helps searchers overcome the challenge of data input on mobile devices, by providing a speech-to-text interface for their mobile search app.

Lots of search systems also leverage the location of the device, with Jones et al. using location-sensitive information to help people discover queries, as opposed to discovering local results [86]. Church et al. have also warned against the intuitive mobile SUI design decision to visualise search results on a map [35] by default. Church et al.'s study concluded that a map can significantly slow searchers down in many common mobile scenarios, such as finding timely information and trivia

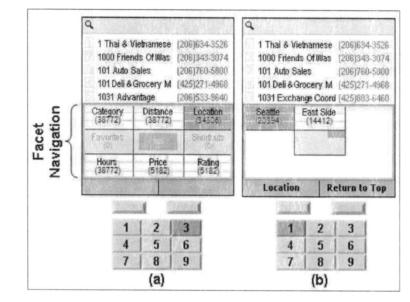

Figure 5.7: FaThumb facet navigation. (a) Pressing 3 navigates to Location. (b) Pressing 1 navigates to Seattle. Image from [94].

(the most common form of mobile information need found by Sohn et al., above). The constant rate of change in mobile technology means that there is a lot of potential in terms of developing novel mobile SUI systems. The recent trend towards building 'apps' for mobile platforms will open up a significant area for innovation in mobile search.

5.5   RE-FINDING, DESKTOP AND PERSONAL SEARCH

Another specialised area of SUIs is to do with re-finding in general, and the searching of one's own information. Beyond searching for files on a computer, Tauscher and Greenberg found that as much as 58% of searching on the Web is to revisit a page [186], while McKenzie and Cockburn saw 81% of web search as re-finding behaviour [124]. In fact, many have shown that re-searching on the Web is much more prevalent than actively bookmarking and revisiting pages [30, 89].

Several of the systems described above were designed for re-finding ones own information, such as Data Mountain (Figure 4.35) and Phlat (Figure 5.6). Much of the work on thumbnails has also focused on people recognising and re-finding results, as were many of the *Personalisable* features discussed in Section 4.4. Further, work by Teevan et al. showed that SUIs can help searchers re-find things by keeping previously visited websites in the same ranked position in a SERP, rather than adjusting them as overall rankings change. Teevan et al. suggested that new results be introduced (if necessary) around previously visited results [188]. Subsequent work by Teevan et al. goes as far as highlighting where a web page being revisiting has changed [191].

Memory and prior context play a significant role in re-finding information both online and within the files of a computer. Early work suggested that a computers files should be organised as a timeline, in a prototype system called Lifestreams [60]. Many have also shown that re-finding information can be difficult (e.g., [20, 189]), with Elsweiler et al., for example, highlighting the notable effect that time has on being able to refind email [55]. Elsweiler designed search systems for email and personal image collections that were specifically informed by the way human memory works [54]. Similarly, Ringel et al. [157] designed a prototype that leveraged episodic memory relating to notable events, which augmented temporally-ordered search results with pictures relating to key events in a searchers calendar and personal collection of photographs. Their results showed that photographic landmarks, as shown in Figure 5.8, significantly reduced document re-finding times.

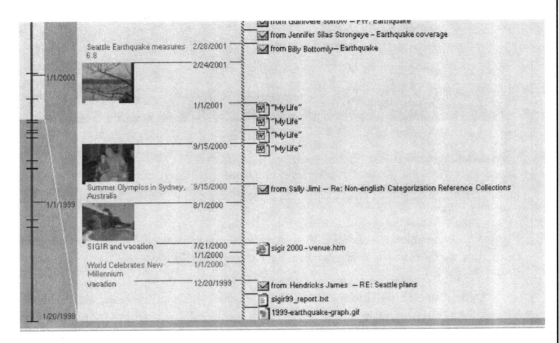

Figure 5.8: Ringel et al. augmented search results with images of landmark points in time, derived from their calendar and photograph collections. Image sampled from [157].

Focusing on more than just temporal contexts, systems like Feldspar (shown in Figure 5.9) [32] and YouPivot [69] try to support memory and context-based recall by helping searchers re-find documents by the associations they have with people, places, events, and other documents.

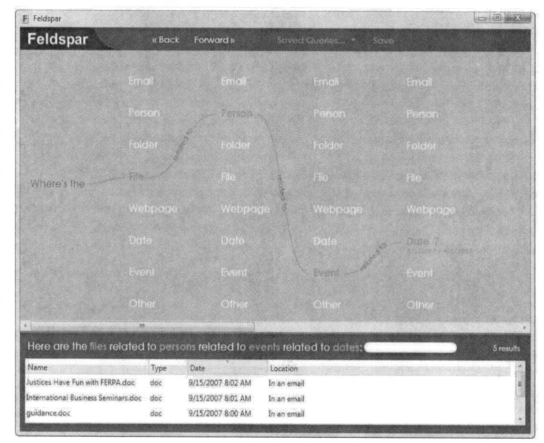

Figure 5.9: Feldspar lets searchers find their own files by letting them build associations across different facets, such as a file from a specific person during a known event. Image from [32].

Desktop and Personal search, given the particular contexts of memory, and the challenging barriers for evaluation created by limitation to personal collections that vary for every possible participant [56], continues to be a rich area for developing research.

5.6 SUMMARY

This chapter has tried to augment the primary content of this book, in Chapter 4, by considering the major novel directions of research and development that diverge from the main body of SUIs that exist. The special factors that relate to mobile searching behaviour, re-finding behaviour, exploratory behaviour, collaborative behaviour, and real-time information, create rich new opportunities for novel SUIs. Each of the areas mentioned, however, as well as specialised domains such as book search [122, 216], structured information retrieval [110], multimedia information retrieval [163], and many others, could be described in their uniqueness in much more detail that can be covered here.

CHAPTER 6

Evaluating Search User Interfaces

A lot can be said about evaluation, both in terms of general HCI evaluation techniques [27], and system-focused IR evaluation techniques [70]. Being only a small part of a book, this section can neither do proper justice to IR evaluation nor HCI evaluation. Consequently, while other related works focus on describing the full range of evaluation methods (e.g., HCI [27] and Interactive IR [70]), this section aims to provide an approach for thinking about SUI evaluation, and what is being evaluated. The aim of this chapter is to help the reader be mindful of approaches and the reasoning for choosing different methods. Readers should then turn to detailed literature specifically about evaluation to better plan their studies. Several approaches, however, are discussed below within the context of the frameworks set out in the introduction of this book.

6.1 IR VS. EMPIRICAL VS. ANALYTICAL APPROACHES

The evaluation of SUIs draws from a mix of experience and disciplines, such as the six introduced at the start of this book, and can thus require a diverse mix of skills. Traditionally, search has been evaluated rather systematically and dogmatically. Interactive IR and the evaluation of human searching, however, has borrowed many methods from disciplines such as Human-Computer Interaction and Social Science. This section focuses on exploring these different approaches in a little more detail, but readers should refer to more significant works in each area.

6.1.1 IR EVALUATION

Traditionally, IR systems are evaluated within TREC-style environments [71]. TREC, which evolved from the original Cranfield experiments [37], allows for different retrieval algorithms to be compared, by providing datasets, specific tasks, and known 'best results' for each task, as calculated by human judges. IR algorithms can then be directly compared for how closely they match the ideal results. Results are evaluated using two key measures: Precision, the proportion of the returned results that are considered to be relevant, and Recall, the proportion of all relevant results that are returned. Average precision, for rewarding result sets with relevant results nearer the top, and Mean Average Precision, for averaging success across all tasks, were used to compare the systems. There are several other tracks within TREC for different types of tasks, such as automatic summarisation of text, question answering, and microblog retrieval[50]. Each of these tracks, however, work independently of a SUI, simply comparing sets of document identifiers for each set task.

Given the success of the TREC approach, an Interactive track was created to take searcher interaction into account. The Interactive Track struggled to find suitable ways of comparing multiple systems directly, as discussed by Belkin and Muresan [18]. Work as early as 1992 had already indicated that simple measures of precision and recall were not sufficient to judge the quality of an IIR system, even with only simple SUIs [184]. Further, the measures of TREC, were based on results being either relevant or not relevant, which cannot be easily applied to IIR systems and broader Web search tasks [22, 23]. Although similar methods, such as Discounted Cumulative Gain [85], and Ranked Half-Life [23] were generated to take relative relevance into account, these measures still did not meet the growing consensus that the quality of result sets alone cannot be used as an appropriate measure of success in interactive forms of IR [18, 22, 184]. Instead, the IIR community moved on to the task-based user study methods used in the HCI domain.

6.1.2   EMPIRICAL USER STUDIES

HCI user study methods instead focus on how well a system, including the SUI, allows searchers to complete a task. It became common to evaluate IIR systems by creating user studies using tasks that are specifically oriented towards search. The success of task-oriented user studies, however, was limited by the quality and appropriateness of the tasks, which led Borland to create a standard for IIR search tasks: the Simulated Work Task [21, 22]. These simulated work tasks are designed to give the participant a scenario and starting search terms, in order to provide participants with a common stance across the five levels of relevance identified by Saracevic [166]. The simulated work tasks, therefore, helps to reduce the variance created by participants and their individual backgrounds. The success of any user study, however, can also be influenced by the motivation of participants, by unforeseen software bugs, and even small user experience differences. As discussed in the introduction, simple graphic design decisions such as choice of colour and layout can affect use of SUIs. Wilson et al. [208] discovered that using two different colours for highlighting created confusion rather than clarity in the design, and chose instead to use faded and non-faded versions of the same colour. Similarly, the presence of small errors or poor UX (such as slow response times) in a SUI can impede participants in taking part in studies and introduce, for example, artificial effects on task performance times. Method decisions, such as paying people to take part in studies, has also been shown to affect the quality of responses [178]. Many of these issues relate to the general design of user studies, and how they are affected by related disciplines such as UX. The introduction of Simulated Work Tasks, however, is an example of applying the established knowledge of Information Seeking and Information Needs to strengthen the use of HCI evaluation techniques for SUIs.

Choosing an appropriate evaluation method for a SUI, given the 6 interrelated disciplines, is difficult. Clustering SUIs, for example, may perform poorly in task-based user studies if the underlying clustering algorithm returns poor clusters. Clustering algorithms, however, had been

---

[50]http://trec.nist.gov/tracks.html

long studied within the TREC framework [41], before they were shown to provide support for searchers in task-based studies [80, 151]. Similarly, Hearst reminds us that the evaluation of faceted browsers is heavily dependent on the quality of metadata generated for the system [77]. Then, given good quality metadata and algorithms, simple UX problems or poorly designed search tasks, can affect the task outcomes or simply user satisfaction and preference. Capra et al, for example, compared two powerful faceted browsers with a standard government website for labor statistics [29]. After quantitative measures found little difference in task performance outcomes, they concluded that powerful but automatic SUI features, like facets, provided comparable support to carefully constructed systems. Qualitative insights further indicated that, despite liking the faceted interaction, the familiar and well-tailored UX provided by the original website was preferable. Essentially, Capra et al. discovered that although automatic and powerful faceted SUI features provide strong support, the overall strength of the UX for the original website was also a crucial factor.

Measurements

Once confident on what is being evaluated, evaluators must then consider what is being measured. Search-based user studies have often led to search-based measures, such as: number of searchers, number of terms per search, number of results visited, search times, task accuracy, and so on. Some have used task-oriented studies to observe and model the Information Seeking process [53], while others have used qualitative methods, such as interviews and observations, to achieve the same [103]. Wilson et al, when studying the use of highlighting in faceted browsers, were evaluating a small UX change (highlighting), and measuring the impact that it had on incidental learning [208]. Although the study methods found significant differences in fact recall between the three versions of their SUI, they found limitations in the UX from the qualitative feedback received in debriefing interviews. Capra et al. used a mixed methods approach to gain both quantitative and qualitative insights into their three SUIs. In the first part of the study, Capra et al. compared the three SUIs using quantitative performance measures, such as time and task accuracy. In the second part, they gained qualitative insights from participants who experienced all three SUIs, which revealed deeper insights into the UX.

There are many measurement and analysis approaches involved in SUI evaluation that are too many to enumerate clearly here. A common approach is to perform log analyses of search queries to understand what people are searching for and how (e.g., [84, 93, 114, 210]). Other studies have used eye tracking technology to study reading patterns or to understand how different SUI features are used during search (e.g., [51, 104, 105]). Instead of tracking the eyes, studies have also focused on mouse movements to understand searching behaviour (e.g., [82, 117]). Many studies also use forms of subjective feedback, such as difficulty [62], usefulness [43], engagement [142], emotion [24], and even cognitive load [73]. Like the design of an evaluation, the choice of measure should also vary significantly towards what the evaluators want to understand and what type of impact the SUI or SUI feature is having.

6.1.3   ANALYTICAL APPROACHES

Most of the methods discussed so far are examples of empirical evaluation methods. Evaluations, especially empirically measured ones, are about observing and recording actual performance. Analytical approaches, on the other hand, are typically low-cost or 'discount' inspection methods [138] that allow evaluators to assess a design and make well informed predictions about the SUI .

There are many analytical methods for UIs and UX, but fewer that have been specifically designed for SUIs. Most analytical methods involve applying established theories to UIs. Heuristic Evaluation [137, 140], for example, provides a formal approach to applying the ten principles listed in Section 2.3 to produce a report on the UX of any UI. Similarly, the Cognitive Walkthrough [201] is a procedure that facilitates UI experts in asking a fixed set of UX questions during each stage of the interaction. Cognitive Walkthroughs presume that a Task Analysis [46] has been done, which breaks down a known set of tasks into a clear set of stages, so that those stages can be applied to UIs; even theoretically if being applied to design concepts and plans. Further, a Cognitive Walkthrough presumes that a set of Personas have been designed, where a Persona is a model of an expected user, such as a novice user and an expert user. Personas allow the evaluators performing the cognitive walkthrough to put themselves in the appropriate mindset or perspective as a novice, which has been shown to improve predictive power of the analysis [155].

There are many other analytical approaches. The Keystroke Level Model (or KLM) [31, 99] method estimated how many interactions it would take to use a UI for a given task. Pre-established timings, such as 0.2s to move a mouse, would then be used to calculate the time taken to perform a task. All of these methods aim to allow the UI designers themselves to critique and assess a design before performing larger, more time-consuming, and more expensive task-based evaluation methods.

Analytical Methods for SUIs

Although search-based variations of evaluation methods are increasingly common, especially with efforts like formalising search tasks with Simulated Work Tasks [21], there are fewer analytical methods specifically for SUIs. Essentially, however, analytical methods are about assessing how well UIs match up with established theories. In lieu of established work in these areas, HCI analytical approaches can be augmented with established theories from Information Science [59]. Task Analyses, for example, can be performed carefully with what we know about the Information Seeking Process [53, 103, 121] and by search tactics [12, 13, 177]. Similarly, Personas and example scenarios could be informed by what we know about different types of search situation [17, 39].

One analytical method that has been specifically designed for SUIs is Sii[51], the Search interface inspector [207, 211]. Sii is similar to the KLM model, in that it allows the designer to analyse how many interactions it takes for searchers to perform specific search tactics; Sii, however, uses higher-level interactions rather than measurable mouse movements. Sii allows designers to step through the different SUI features that have been implemented, one at a time, and ask how many 'moves' it takes to perform each of Bates' 32 search tactics [12, 13]. 'Moves' are defined, as according to Bates,

---

[51]http://mspace.fm/sii

as a physical and mental actions that are taken to perform a tactic [14], such as making a choice, entering a search query, clicking on a result, etc. Further, Wilson et al. [211] created a model to connect Bates' tactics with Belkin et al.'s different searcher conditions [17]. This connection allows the evaluator, using Sii, to assess the suitability of a SUI for different searcher conditions, as well as total support for different types of search tactics, and the strength of how different SUI features have been designed.

The analytical methods described above have, like any model-driven predictive approach, one major limitation: they can only make estimates about how suitable a design is, or help to refine a UX before it is formally evaluated. Such analytical methods, however, can help make assessments of different SUI features to help people make better designs in the first place. The key factor to remember about analytical approaches is that they can utilise the wealth of theoretical models such that our designs can leverage what we already know about information seeking, behaviour, and human searchers.

6.2   CHOOSING AN APPROACH

The most important thing to consider, when planning an evaluation of a SUI, is which aspect is being evaluated and what type of contribution it is making. The aim of this book, and this section so far, has been to highlight the importance of thinking about SUIs within the same framework set out in Chapter 2. This framework is equally as important for evaluation as it is for understanding SUIs. Consequently, to evaluate a SUI, the evaluator must be confident in where their SUI is innovating (*Input, Control, Informational,* or *Personalisable* features) and whether their changes are primarily related to the SUI, the UX, the IR algorithms, and so on. The evaluation of SUIs, however, is as intricately influenced by the 6 disciplines that affect SUIs themselves (shown in Figure 2.2). If innovating algorithmically, for example, then purely IR evaluation approaches might be most appropriate. If an evaluator thinks these changes will also then affect search success or efficiency, then empirical evaluations with human participation is needed. Further, if changes are product and client driven, then evaluations with target users may be required to streamline and perfect a final UX. Thinking about evaluation in this way should help evaluators to choose methods and measures from the most closely related disciplines to their innovations, while being aware of how they relate to the other methods.

One approach to choosing an evaluation approach, provided by the HCI community, is the DECIDE process [154]. There are six parts to this process:

- D – Determine the goals of the evaluation. What do you want to prove or examine?

- E – Explore the specific questions to be answered. Which element of the framework is being effected? Which disciplines are involved?

- C – Choose an evaluation paradigm, such as systematic IR or empirical user studies.

- I – Identify the practical issues in performing such an evaluation, such as structured tasks and appropriate datasets.

- D – Decide how to deal with any ethical issues. Ethical issues are especially important when working with humans.

- E – Evaluate, Interpret, and present the data.

This DECIDE process begins with identifying where results have innovated and which types of disciplines are involved in the SUI feature that is being evaluated. Consequently, the approach is good for making sure that the right questions have been asked about an evaluation. It may be that this DECIDE approach leads to a traditional and systematic IR evaluation, or to embark upon an exploratory and creative design process. Regardless, however, it is the stages of decision that help to clarify exactly which evaluation method is most suitable to the SUI or SUI feature that has been created. Once decided, however, one recommendation that is common to all evaluation approaches, is to perform a pilot study to make sure that both the evaluation and the evaluator are properly prepared to perform it correctly.

6.3 SUMMARY

The main aim of this book has been to provide readers with a framework for thinking about SUI features, how they support searchers, and how they could be extended to provide better or broader support in different areas. Consequently, this section has also highlighted that evaluation should be considered in the same way. While thinking analytically about what type of support a SUI feature is providing, analytical methods can be applied to assess their strength and their purpose. Similarly, understanding the kind of support that a SUI feature is providing, helps us to better estimate how they should be evaluated, and using which types of methods. Evaluating SUIs, therefore, requires a mix of expertise from both the IR an HCI domains, which is why more extensive books have been written purely about this topic.

CHAPTER 7

Conclusions

This book has focused on describing Search User Interfaces (SUIs) and SUI features. To do this, the book began by breaking down search features, using the current Google SUI as an example, into four groups: *Input, Control, Informational,* and *Personalisable* features. The book then introduced how we should think about SUIs, in terms of the people who will use them, and the technology that enables them. In summary, we should think about the ways in which we want to support searchers with a SUI, and then find the technology that enables us to do so. Further, the book described six different disciplines that are involved in designing and building SUIs, recommending that we think of each SUI feature in terms of the contribution made by or for each discipline.

The book proceeded to describe examples of the four categories of SUI features. *Input* features enable searchers in describing what they are searching for. Beyond searching with keywords on the web, most approaches strive to provide metadata to searchers that can be recognised and selected. *Control* features help us to manipulate our searches, either by refining what we are looking for, or by filtering the results to be more specific. *Informational* features provide results to searchers in many different forms, but can also help the searcher be aware of what they have done and can do next. Finally, *Personalisable* features are designed to impact the preceding types of features so that they mean more to the individual searcher.

Finally, this book reviewed the ways in which we should evaluate SUIs and their features, especially using techniques that are built upon established theories of how people search.

7.1 REVISITING HOW WE THINK ABOUT SUIs

UIs, and thus SUIs too, are designed to help some user, or searcher, to achieve some goal. Consequently, we should think about SUIs in terms of how they help people and how they can be strengthened to support searchers better. While it is possible to influence or change human behaviour with a SUI, data and technology are under our control and can be adjusted to suit the way we want to support searchers. Further, to make any change in how we support searchers, SUI designers need to consider how it affects and can be further affected by each of the six disciplines.

The design approaches for any UI has been well studied by the HCI community, which has provided many general design recommendations, such as those described in Section 2.3. Similarly, the there are UX specialists who can design, and more importantly, focus on the finished tailored user experiences of specific products and systems. There are also graphic designers who work with the visual aesthetics of a system, such as brand and use of colour.

There are also several search specific disciplines, such as Library and Information Science, Information Retrieval and the Information Seeking communities. Underneath any SUI, IR algorithms process search requests, analyse datasets, and return results. These two are dependent on each other, the SUI to allow the algorithms to be used, and the algorithms to allow the SUI to be useful. Library and Information Science have studied the design, storage, and use of information for much longer than there have been computers, and inform much about the quality of information that can be accessed through SUIs. Finally, the Information Seeking community can perhaps tell us most about how the systems, the data, the SUIs, and the ideal user experience. It is the theories of Information Seeking that can tell us about the process people go through, and the types of searches that people may perform. This last community is perhaps the best place to start when thinking about supporting searchers with a new SUI, as novel interactions, types of metadata, and algorithms can be designed to support the different ways in which we know people search.

7.2 INNOVATING AND EVALUATING SUIs

Chapter 4 of this book reviewed many SUI innovations by framing them into four major categories. SUIs tend to innovate in one of these areas, whether providing new types of faceted metadata to increase the power of *Input* and *Control,* or whether it is new *Informational* visualisations to display and explore the results. Many of the stronger innovations act in multiple areas. Even the standard *Informational* display of search results was strengthened to display usable links, as *Input* and *Control,* and even *Personalisable* information about searchers' previous actions and the actions of other searchers. Similarly, work in faceted browsers explored turning passive columns (ready for *Input and Control*) to be *Pesronalisable* representations that highlight previous actions and *Informational* visualisations that guide searchers towards new information.

Further, SUI features can be considered by how they are innovating in these areas, where simple UX changes, like highlights for example, can provide significant additional power to an already innovative SUI feature. Similarly, we can consider how new IR algorithms, or new forms of metadata, can be used by SUIs to allow searchers to change the way they *Input* queries or *Control* and filter results. This book aimed to provide readers with a way of thinking about how SUIs and SUI features are innovating. This framework may help to first consider new innovations, and then how to effectively and appropriately evaluate them.

7.3 SUMMARY OF DESIGN RECOMMENDATIONS

In describing these different types of features and the SUIs they have been included in, several SUI-specific design recommendations have been provided that extend the general design considerations that were introduced in Section 2.3. These SUI-specific recommendations are:

1. Keep the search box and the current query clearly visible for the searcher at all times.

2. Help searchers to create useful queries whenever possible.

3. Make it clear how results relate to metadata in your system, to help searchers to judge the results and make sense of the whole collection.

4. Carefully curated metadata is better than automatically generated, but both are better than no metadata at all.

5. If appropriate, support searchers in reviewing their decisions and their options quickly and easily.

6. Always return results based on the first interaction, as subsequent interactions may never be needed.

7. Never let searchers reach a dead end, where they have to go back or start over.

8. Help searchers avoid mistakes wherever possible, but do not force that help upon them.

9. Give searchers control over the way results are ordered.

10. Make sure it is obvious exactly how results are ordered and which, therefore, are most important.

11. Avoid unnecessary information, which can be distracting during search.

12. Searchers rarely scroll, so get 'important' information above the first-scroll point.

13. Provide actionable features in the SERP results directly so that searchers do not have to interrupt their search.

14. Images and Previews can help searchers make better browsing decisions.

15. Make sure the dimensions and layout of a visualisation are clear and intuitive to the searcher.

16. Guiding numbers help searchers to make better browsing decisions

17. Animation should be used carefully and purposefully to convey a message, such as change.

18. Track and reuse information about the behaviour of a system's searchers.

19. Help searchers to return to previously viewed SERPs and results.

20. Help searchers to recover their previous search sessions, as they may be back to finish a task.

These recommendations should be considered in the context of more general UI and UX recommendations, such as those considered in Section 2.3.

7.4   CONCLUDING REMARKS

If this book inspired you, as a reader, to think more about SUIs, there are other notable surveys that have been provided by both academia and industry (e.g., [76, 132, 209]). The goal of this book, however, has been to provide a framework for thinking about SUI designs, their contributions, and how they should be evaluated, while populating that framework with a comprehensive review of previous literature. Readers should think about the searchers first, and then the technologies that will help design the right SUI. When assessing or evaluating existing SUIs, readers should think about the different features, and the one or more categories they fall into, in order to identify what they want to learn about the SUI feature and thus how to design an evaluation. When designing new SUIs, readers should think about the range of support they are providing with the combination of features being included. Further, readers should consider whether features could be extended or improved so that they fall into more than one category of feature. Examining and understanding SUIs and SUI features in this way leads us to be more aware of how and why some SUIs are successful, and whether new ideas can have an impact too.

Bibliography

[1] Ahlberg, C. and Shneiderman, B., Visual information seeking: tight coupling of dynamic query filters with starfield displays. In *Proceedings of the SIGCHI conference on Human factors in computing systems*, ACM, Boston, Massachusetts, United States, 313–317, 1994. DOI: 10.1145/191666.191775 Cited on page(s) 47

[2] Ahlberg, C., Williamson, C. and Shneiderman, B., Dynamic queries for information exploration: an implementation and evaluation. In *Proceedings of the SIGCHI conference on Human factors in computing systems*, ACM, Monterey, California, United States, 619–626, 1992. DOI: 10.1145/142750.143054 Cited on page(s) 47

[3] Amento, B., Terveen, L., Hill, W. and Hix, D., TopicShop: enhanced support for evaluating and organizing collections of Web sites. In *Proceedings of the 13th annual ACM symposium on User interface software and technology*, ACM, San Diego, California, United States, 201–209, 2000. DOI: 10.1145/354401.354771 Cited on page(s) 62, 65

[4] Amershi, S., Fogarty, J., Kapoor, A. and Tan, D., Overview based example selection in end user interactive concept learning. In *Proceedings of the 22nd annual ACM symposium on User interface software and technology*, ACM, Victoria, BC, Canada, 247–256, 2009. DOI: 10.1145/1622176.1622222 Cited on page(s) 32

[5] Amershi, S. and Morris, M.R., CoSearch: a system for co-located collaborative web search. In *Proceeding of the twenty-sixth annual SIGCHI conference on Human factors in computing systems*, ACM, Florence, Italy, 1647–1656, 2008. DOI: 10.1145/1357054.1357311 Cited on page(s) 82

[6] André, P., Wilson, M.L., Russell, A., Smith, D.A. and Owens, A., Continuum: designing timelines for hierarchies, relationships and scale. In *Proceedings of the 20th annual ACM symposium on User interface software and technology*, ACM, Newport, RI, USA, 101–110, 2007. DOI: 10.1145/1294211.1294229 Cited on page(s) 64, 68

[7] Anick, P.G., Brennan, J.D., Flynn, R.A., Hanssen, D.R., Alvey, B. and Robbins, J.M., A direct manipulation interface for boolean information retrieval via natural language query. In *Proceedings of the 13th annual international ACM SIGIR conference on Research and development in information retrieval*, ACM, Brussels, Belgium, 135–150, 1990. DOI: 10.1145/96749.98015 Cited on page(s) 21, 23

[8] Back, J. and Oppenheim, C., A model of cognitive load for IR: implications for user relevance feedback interaction. *Information Research*, 6(2), 2001. Cited on page(s) 27

[9] Baeza-Yates, R. and Riberio-Neto, B., Modern Information Retrieval. Addison-Wesley Publishing Company, 2008. Cited on page(s) 2

[10] Bates, M.J., The cascade of interactions in the digital library interface. *Information Processing & Management*, 38(3), 381–400, 2002. DOI: 10.1016/S0306-4573(01)00041-3   Cited on page(s) 9

[11] Bates, M.J., The Design of Browsing and Berrypicking Techniques for the Online Search Interface. *Online Information Review*, 13(5), 407–424, 1989. DOI: 10.1108/eb024320 Cited on page(s) 2

[12] Bates, M.J., Idea tactics. *Journal of the American Society for Information Science*, 30(5), 280–289, 1979. DOI: 10.1002/asi.4630300507 Cited on page(s) 12, 96

[13] Bates, M.J., Information search tactics. *Journal of the American Society for Information Science*, 30(4), 205–214, 1979. DOI: 10.1002/asi.4630300406 Cited on page(s) 12, 29, 96

[14] Bates, M.J., Where should the person stop and the information search interface start? *Information Processing & Management*, 26(5), 575–591, 1990.
DOI: 10.1016/0306-4573(90)90103-9 Cited on page(s) 97

[15] Baudisch, P., Tan, D., Collomb, M., Robbins, D., Hinckley, K., Agrawala, M., Zhao, S. and Ramos, G., Phosphor: explaining transitions in the user interface using afterglow effects. In *Proceedings of the 19th annual ACM symposium on User Interface Software and Technology*, ACM, Montreux, Switzerland, 169–178, 2006. DOI: 10.1145/1166253.1166280 Cited on page(s) 11, 73

[16] Belkin, N.J., Cool, C., Stein, A. and Thiel, U., Cases, scripts, and information-seeking strategies: on the design of interactive information retrieval systems. *Expert Systems with Applications*, 9(3), 379–395, 1995. DOI: 10.1016/0957-4174(95)00011-W Cited on page(s) 18, 19

[17] Belkin, N.J., Marchetti, P.G. and Cool, C., Braque: design of an interface to support user interaction in information retrieval. *Information Processing & Management*, 29(3), 325–344, 1993. DOI: 10.1016/0306-4573(93)90059-M Cited on page(s) 12, 96, 97

[18] Belkin, N.J. and Muresan, G., Measuring web search effectiveness: Rutgers at Interactive TREC. In *WWW2004 Workshop on Measuring Web Search Effectiveness: The User Perspective*, New York, NY, USA, 2004. Cited on page(s) 94

[19] Bernstein, M.S., Suh, B., Hong, L., Chen, J., Kairam, S. and Chi, E.H., Eddi: interactive topic-based browsing of social status streams. In *Proceedings of the 23nd annual ACM symposium on User interface software and technology*, ACM, New York, New York, USA, 303–312, 2010. DOI: 10.1145/1866029.1866077 Cited on page(s) 84, 85

[20] Boardman, R. and Sasse, M.A., "Stuff goes into the computer and doesn't come out": a cross-tool study of personal information management. In *Proceedings of the SIGCHI conference on Human factors in computing systems*, ACM, Vienna, Austria, 583–590, 2004. DOI: 10.1145/985692.985766 Cited on page(s) 89

[21] Borlund, P., Experimental Components for the Evaluation of Interactive Information Retrieval Systems. *Journal of Documentation*, 56(1), 71–90, 2000. DOI: 10.1108/EUM0000000007110 Cited on page(s) 94, 96

[22] Borlund, P., The IIR evaluation model: a framework for evaluation of interactive information retrieval systems. *Information Research*, 8(3), Paper 152, 2003. DOI: 10.1108/eb026574 Cited on page(s) 94

[23] Borlund, P. and Ingwersen, P., Measures of relative relevance and ranked half-life: performance indicators for interactive IR. In *Proceedings of the 21st annual international ACM SIGIR conference on Research and development in information retrieval*, ACM, Melbourne, Australia, 324–331, 1998. DOI: 10.1145/290941.291019 Cited on page(s) 94

[24] Bradley, M.M. and Lang, P.J., Measuring emotion: The self-assessment manikin and the semantic differential. *Journal of Behavior Therapy and Experimental Psychiatry*, 25(1), 49–59, 1994. DOI: 10.1016/0005-7916(94)90063-9 Cited on page(s) 95

[25] Brin, S. and Page, L., The anatomy of a large-scale hypertextual Web search engine. *Computer Networks and ISDN Systems*, 30(1–7), 107-117, 1998. DOI: 10.1016/S0169-7552(98)00110-X Cited on page(s) 13, 20

[26] Buckley, C., Salton, G. and Singhal, A., Automatic Query Expansion using SMART: TREC-3. In *Proceedings of the Third Text REtrieval Conference (TREC-3)*, NIST Special Publication, 500–525, 1995. Cited on page(s) 22

[27] Cairns, P. and Cox, A.L., Research Methods for Human-Computer Interaction. Cambridge University Press, 2008. Cited on page(s) 93

[28] Capone, L., Competitive Analysis of Ask Jeeves Search Engines, VeriTest Inc, 84.2004. Cited on page(s) 54

[29] Capra, R., Marchionini, G., Oh, J.S., Stutzman, F. and Zhang, Y., Effects of structure and interaction style on distinct search tasks. In *Proceedings of the 7th ACM/IEEE-CS joint conference on Digital libraries*, ACM, Vancouver, BC, Canada, 442–451, 2007. DOI: 10.1145/1255175.1255267 Cited on page(s) 95

[30] Capra, R.G. and Perez-Quinones, M.A., Using Web Search Engines to Find and Refind Information. *Computer*, 38(10), 36–42, 2005. DOI: 10.1109/MC.2005.355 Cited on page(s) 13, 88

[31] Card, S.K., Moran, T.P. and Newell, A., The keystroke-level model for user performance time with interactive systems. *Communications of the ACM*, 23(7), 396–410, 1980. DOI: 10.1145/358886.358895 Cited on page(s) 96

[32] Chau, D.H., Myers, B. and Faulring, A., What to do when search fails: finding information by association. In *Proceeding of the twenty-sixth annual SIGCHI conference on Human factors in computing systems*, ACM, Florence, Italy, 999–1008, 2008. DOI: 10.1145/1357054.1357208 Cited on page(s) 89, 90

[33] Chen, M., Hearsty, M., Hong, J. and Lin, J., Cha-Cha: a system for organizing intranet search results. In *Proceedings of the 2nd conference on USENIX Symposium on Internet Technologies and Systems - Volume 2*, USENIX Association, Boulder, Colorado, 5–5, 1999. Cited on page(s) 33

[34] Chilton, L.B. and Teevan, J., Addressing people's information needs directly in a web search result page. In *Proceedings of the 20th international conference on World wide web*, ACM, Hyderabad, India, 27–36, 2011. DOI: 10.1145/1963405.1963413 Cited on page(s) 70

[35] Church, K., Neumann, J., Cherubini, M. and Oliver, N., The "Map Trap"?: an evaluation of map versus text-based interfaces for location-based mobile search services. In *Proceedings of the 19th international conference on World wide web*, ACM, Raleigh, North Carolina, USA, 261–270, 2010. DOI: 10.1145/1772690.1772718 Cited on page(s) 87

[36] Clarkson, E.C., Navathe, S.B. and Foley, J.D., Generalized formal models for faceted user interfaces. In *Proceedings of the 9th ACM/IEEE-CS joint conference on Digital libraries*, ACM, Austin, TX, USA, 125–134, 2009. DOI: 10.1145/1555400.1555422 Cited on page(s) 40

[37] Cleverdon, C.W., The Cranfield Tests on Index Language Devices. *Aslib Proceedings*, 19(6), 173–194, 1967. DOI: 10.1108/eb050097 Cited on page(s) 2, 93

[38] Cockburn, A. and McKenzie, B., An evaluation of cone trees. In *Proceedings of the British Computer Society Conference on Human-Computer Interaction*, Springer-Verlag, Sunderland, UK, 425–436, 2000. Cited on page(s) 66

[39] Cool, C. and Belkin, N.J., A classification of interactions with information. In *Proceedings of the Fourth International Conference on Conceptions of Library and Information Science (CoLIS 4)*, Lirbaries Unlimited, Seattle, WA, USA, 1–15, 2002. Cited on page(s) 96

[40] Cutrell, E., Robbins, D., Dumais, S. and Sarin, R., Fast, flexible filtering with phlat. In *Proceedings of the SIGCHI conference on Human Factors in computing systems*, ACM, Montréal, Québec, Canada, 261–270, 2006. DOI: 10.1145/1124772.1124812 Cited on page(s) 86

[41] Cutting, D.R., Karger, D.R., Pedersen, J.O. and Tukey, J.W., Scatter/Gather: a cluster-based approach to browsing large document collections. In *Proceedings of the 15th annual international ACM SIGIR conference on Research and development in information retrieval*, ACM, Copenhagen, Denmark, 318–329, 1992. DOI: 10.1145/133160.133214 Cited on page(s) 10, 13, 95

[42] Cyr, D., Head, M. and Larios, H., Colour appeal in website design within and across cultures: A multi-method evaluation. *International Journal of Human-Computer Studies*, 68(1–2), 1-21, 2010. DOI: 10.1016/j.ijhcs.2009.08.005 Cited on page(s) 11

[43] Davis, F.D., Perceived usefulness, perceived ease of use, and user acceptance of information technology. *MIS Quarterly*, 13(3), 319–340, 1989. DOI: 10.2307/249008 Cited on page(s) 95

[44] Dervin, B., From the mindís eye of the user: The sense-making qualitative-quantitative methodology. in Glazier, J.D. and Powell, R.R. eds. *Qualitative research in information management*, Englewood, CO: Libraries Unlimited Inc, 1992, 61–84. Cited on page(s) 2, 84

[45] Dervin, B., An overview of sense-making research: concepts, methods and results to date. In *Proceedings of the International Communications Association Annual Meeting*, Dallas Texas, 1983. Cited on page(s) 2, 84

[46] Diaper, D. and Stanton, N., The handbook of task analysis for human-computer interaction. CRC, 2003. Cited on page(s) 11, 13, 96

[47] Diriye, A., Blandford, A. and Tombros, A., A polyrepresentational approach to interactive query expansion. In *Proceedings of the 9th ACM/IEEE-CS joint conference on Digital libraries*, ACM, Austin, TX, USA, 217–220, 2009. DOI: 10.1145/1555400.1555434 Cited on page(s) 41

[48] Doan, K., Plaisant, C. and Shneiderman, B., Query Previews in Networked Information Systems. In *Proceedings of the 3rd International Forum on Research and Technology Advances in Digital Libraries*, IEEE Computer Society, 120, 1996. DOI: 10.1109/ADL.1996.502522 Cited on page(s) 68

[49] Drori, O. and Alon, N., Using documents classification for displaying search results list. *Journal of Information Science*, 29(2), 97–106, 2003. DOI: 10.1177/016555150302900202 Cited on page(s) 34

[50] Dumais, S., Cutrell, E. and Chen, H., Optimizing search by showing results in context. In *Proceedings of the SIGCHI conference on Human factors in computing systems*, ACM, New York, NY, USA, 277–284, 2001. DOI: 10.1145/365024.365116 Cited on page(s) 33

[51] Dumais, S.T., Buscher, G. and Cutrell, E., Individual differences in gaze patterns for web search. In *Proceeding of the third symposium on Information interaction in context*, ACM, New Brunswick, New Jersey, USA, 185–194, 2010. DOI: 10.1145/1840784.1840812   Cited on page(s) 95

[52] Egan, D.E., Remde, J.R., Gomez, L.M., Landauer, T.K., Eberhardt, J. and Lochbaum, C.C., Formative design evaluation of superbook. *ACM Transactions on Information Systems*, 7(1), 30–57, 1989. DOI: 10.1145/64789.64790 Cited on page(s) 33

[53] Ellis, D., A behavioural approach to information retrieval system design. *Journal of Documentation*, 45(3), 171–212, 1989. DOI: 10.1108/eb026843 Cited on page(s) 2, 12, 95, 96

[54] Elsweiler, D., Supporting Human Memory in Personal Information Management *Department of Computer and Information Sciences*, University of Strathclyde, 298.2007. DOI: 10.1145/1394251.1394270 Cited on page(s) 89

[55] Elsweiler, D., Baillie, M. and Ruthven, I., What Makes Re-finding Information Difficult? A Study of Email Re-finding. in Clough, P., Foley, C., Gurrin, C., Jones, G., Kraaij, W., Lee, H. and Mudoch, V. eds. *Advances in Information Retrieval*, Springer Berlin / Heidelberg, 2011, 568–579. Cited on page(s) 89

[56] Elsweiler, D. and Ruthven, I., Towards task-based personal information management evaluations. In *Proceedings of the 30th annual international ACM SIGIR conference on Research and development in information retrieval*, ACM, Amsterdam, The Netherlands, 23–30, 2007. DOI: 10.1145/1277741.1277748 Cited on page(s) 90

[57] Elsweiler, D., Wilson, M.L. and Lunn, B.K., Understanding Casual-leisure Information Behaviour. in Spink, A. and Heinstrom, J. eds. *New Directions in Information Behaviour*, Emerald, 2011, (in press). Cited on page(s) 87

[58] Evans, B.M. and Chi, E.H., Towards a model of understanding social search. In *Proceedings of the 2008 ACM conference on Computer supported cooperative work*, ACM, San Diego, CA, USA, 485–494, 2008. DOI: 10.1145/1460563.1460641 Cited on page(s) 82

[59] Fisher, K.E., Erdelez, S. and Mckechnie, L., Theories of information behavior. Information Today Inc, 2005. Cited on page(s) 12, 96

[60] Freeman, E. and Gelernter, D., Lifestreams: a storage model for personal data. *ACM SIGMOD Record*, 25(1), 80–86, 1996. DOI: 10.1145/381854.381893 Cited on page(s) 89

[61] Geller, V.J. and Lesk, M.E., User interfaces to information systems: choices vs. commands. In *Proceedings of the 6th annual international ACM SIGIR conference on Research and development in information retrieval*, ACM, Bethesda, Maryland, 130–135, 1983. DOI: 10.1145/511793.511813 Cited on page(s) 19

[62] Ghani, J.A., Supnick, R. and Rooney, P., The experience of flow in computer-mediated and in face-to-face groups. In *Proceedings of the twelfth international conference on Information systems*, University of Minnesota, New York, New York, United States, 229–237, 1991. Cited on page(s) 95

[63] Ghias, A., Logan, J., Chamberlin, D. and Smith, B.C., Query by humming: musical information retrieval in an audio database. In *Proceedings of the third ACM international conference on Multimedia*, ACM, San Francisco, California, USA, 231–236, 1995. DOI: 10.1145/217279.215273 Cited on page(s) 32

[64] Golbeck, J. and Hendler, J., FilmTrust: movie recommendations using trust in web-based social networks. In *Proceedings of the 3rd IEEE Consumer Communications and Networking Conference, 2006*, IEEE, Las Vegas, NV, USA, 282–286, 2006. DOI: 10.1109/CCNC.2006.1593032 Cited on page(s) 77

[65] Golovchinsky, G., Qvarfordt, P. and Pickens, J., Collaborative Information Seeking. *Computer*, 42(3), 47–51, 2009. DOI: 10.1109/MC.2009.73 Cited on page(s) 82

[66] Gwizdka, J., Distribution of cognitive load in Web search. *Journal of the American Society for Information Science and Technology*, 61(11), 2167–2187, 2010. DOI: 10.1002/asi.21385 Cited on page(s) 34

[67] Gwizdka, J., What a difference a tag cloud makes: Effects of tasks and cognitive abilities on search results interface use. *Information Research*, 14(4), 2009. Cited on page(s) 40

[68] Haas, K., Mika, P., Tarjan, P. and Blanco, R., Enhanced results for web search. In *Proceedings of the 34th international ACM SIGIR conference on Research and development in Information*, ACM, Beijing, China, 725–734, 2011. Cited on page(s) 50

[69] Hailpern, J., Jitkoff, N., Warr, A., Karahalios, K., Sesek, R. and Shkrob, N., YouPivot: improving recall with contextual search. In *Proceedings of the 2011 annual conference on Human factors in computing systems*, ACM, Vancouver, BC, Canada, 1521–1530, 2011. DOI: 10.1145/1978942.1979165 Cited on page(s) 89

[70] Harman, D., Information Retrieval Evaluation. *Synthesis Lectures on Information Concepts, Retrieval, and Services*, 3(2), 1–119, 2011. DOI: 10.2200/S00368ED1V01Y201105ICR019 Cited on page(s) 93

[71] Harman, D., Overview of the first TREC conference. In *Proceedings of the 16th annual international ACM SIGIR conference on Research and development in information retrieval*, ACM, Pittsburgh, Pennsylvania, United States, 36–47, 1993. DOI: 10.1145/160688.160692 Cited on page(s) 2, 93

[72] Harman, D., Towards interactive query expansion. In *Proceedings of the 11th annual international ACM SIGIR conference on Research and development in information retrieval*, ACM, Grenoble, France, 321–331, 1988. DOI: 10.1145/62437.62469 Cited on page(s) 27

[73] Hart, S.G. and Staveland, L.E., Development of NASA-TLX (Task Load Index): Results of Empirical and Theoretical Research. in Peter, A.H. and Najmedin, M. eds. *Advances in Psychology*, North-Holland, 1988, 139–183. Cited on page(s) 95

[74] Hassenzahl, M. and Tractinsky, N., User experience: a research agenda. *Behaviour & Information Technology*, 25(2), 91–97, 2006. DOI: 10.1080/01449290500330331 Cited on page(s) 11

[75] Hearst, M., Design Recommendations for Hierarchical Faceted Search Interfaces. In *SIGIR2006 Workshop on Faceted Search*, Seattle, Washington, USA, 2006. Cited on page(s) 11, 40

[76] Hearst, M., Search User Interfaces. Cambridge University Press, 2009. Cited on page(s) 102

[77] Hearst, M.A., Clustering versus faceted categories for information exploration. *Communications of the ACM*, 49(4), 59–61, 2006. DOI: 10.1145/1121949.1121983 Cited on page(s) 35, 95

[78] Hearst, M.A., TileBars: visualization of term distribution information in full text information access. In *Proceedings of the SIGCHI conference on Human factors in computing systems*, ACM Press/Addison-Wesley Publishing Co., Denver, Colorado, United States, 59–66, 1995. DOI: 10.1145/223904.223912 Cited on page(s) 58, 59

[79] Hearst, M.A. and Karadi, C., Cat-a-Cone: an interactive interface for specifying searches and viewing retrieval results using a large category hierarchy. In *Proceedings of the 20th annual international ACM SIGIR conference on Research and development in information retrieval*, ACM, Philadelphia, Pennsylvania, United States, 246–255, 1997. DOI: 10.1145/258525.258582 Cited on page(s) 65

[80] Hearst, M.A. and Pedersen, J.O., Reexamining the cluster hypothesis: scatter/gather on retrieval results. In *Proceedings of the 19th annual international ACM SIGIR conference on Research and development in information retrieval*, ACM, Zurich, Switzerland, 76–84, 1996. DOI: 10.1145/243199.243216 Cited on page(s) 34, 35, 95

[81] Hoeber, O. and Yang, X.D., HotMap: Supporting visual exploration of Web search results. *Journal of the American Society for Information Science and Technology*, 60(1), 90–110, 2009. DOI: 10.1002/asi.20957 Cited on page(s) 58

[82] Huang, J., White, R.W. and Dumais, S., No clicks, no problem: using cursor movements to understand and improve search. In *Proceedings of the 2011 annual conference on*

Human factors in computing systems, ACM, Vancouver, BC, Canada, 1225–1234, 2011. DOI: 10.1145/1978942.1979125 Cited on page(s) 95

[83] Hurlock, J. and Wilson, M.L., Searching Twitter: Separating the Tweet from the Chaff. In *Proceedings of the Fifth International AAAI Conference on Weblogs and Social Media*, AAAI, Barcelona, Spain, 2011. Cited on page(s) 84

[84] Jansen, B.J., Spink, A. and Saracevic, T., Real life, real users, and real needs: a study and analysis of user queries on the web. *Information Processing & Management*, 36(2), 207–227, 2000. DOI: 10.1016/S0306-4573(99)00056-4 Cited on page(s) 30, 95

[85] Järvelin, K. and Kekäläinen, J., IR evaluation methods for retrieving highly relevant documents. In *Proceedings of the 23rd annual international ACM SIGIR conference on Research and development in information retrieval*, ACM, Athens, Greece, 41–48, 2000. DOI: 10.1145/345508.345545 Cited on page(s) 94

[86] Jones, M., Buchanan, G., Harper, R. and Xech, P.-L., *Questions* not *answers*: a novel mobile search technique. In *Proceedings of the SIGCHI conference on Human factors in computing systems*, ACM, San Jose, California, USA, 155–158, 2007. DOI: 10.1145/1240624.1240648 Cited on page(s) 87

[87] Jones, M. and Marsden, G., Mobile interaction design. Wiley, 2006. Cited on page(s) 87

[88] Jones, S., A statistical interpretation of term specificity and its application in retrieval. *Journal of Documentation*, 28(1), 11–20, 1972. DOI: 10.1108/eb026526 Cited on page(s) 2, 22

[89] Jones, W., Bruce, H. and Dumais, S., Keeping found things found on the web. In *Proceedings of the tenth international conference on Information and knowledge management*, ACM, Atlanta, Georgia, USA, 119–126, 2001. DOI: 10.1145/502585.502607 Cited on page(s) 88

[90] Kaasten, S., Greenberg, S. and Edwards, C., How people recognize previously seen Web pages from titles, URLs and thumbnails. In *Proceedings of Human Computer Interaction*, 247–265, 2002. DOI: 10.1145/1402256.1402258 Cited on page(s) 54

[91] Kammerer, Y., Nairn, R., Pirolli, P. and Chi, E.H., Signpost from the masses: learning effects in an exploratory social tag search browser. In *Proceedings of the 27th international conference on Human factors in computing systems*, ACM, Boston, MA, USA, 625–634, 2009. DOI: 10.1145/1518701.1518797 Cited on page(s) 11, 41

[92] Kamvar, M. and Baluja, S., A large scale study of wireless search behavior: Google mobile search. In *Proceedings of the SIGCHI conference on Human Factors in computing systems*, ACM, Montréal, Québec, Canada, 701–709, 2006. DOI: 10.1145/1124772.1124877 Cited on page(s) 87

[93] Kamvar, M., Kellar, M., Patel, R. and Xu, Y., Computers and iphones and mobile phones, oh my!: a logs-based comparison of search users on different devices. In *Proceedings of the 18th international conference on World wide web*, ACM, Madrid, Spain, 801–810, 2009. DOI: 10.1145/1526709.1526817 Cited on page(s) 30, 87, 95

[94] Karlson, A.K., Robertson, G.G., Robbins, D.C., Czerwinski, M.P. and Smith, G.R., FaThumb: a facet-based interface for mobile search. In *Proceedings of the SIGCHI conference on Human Factors in computing systems*, ACM, Montréal, Québec, Canada, 711–720, 2006. DOI: 10.1145/1124772.1124878 Cited on page(s) 87, 88

[95] Kay, B.M. and Watters, C., Exploring multi-session web tasks. In *Proceeding of the twenty-sixth annual SIGCHI conference on Human factors in computing systems*, ACM, Florence, Italy, 1187–1196, 2008. DOI: 10.1145/1357054.1357243 Cited on page(s) 76

[96] Kelly, D., Cushing, A., Dostert, M., Niu, X. and Gyllstrom, K., Effects of popularity and quality on the usage of query suggestions during information search. In *Proceedings of the 28th international conference on Human factors in computing systems*, ACM, Atlanta, Georgia, USA, 45–54, 2010. DOI: 10.1145/1753326.1753334 Cited on page(s) 27

[97] Kerne, A., Koh, E., Smith, S.M., Webb, A. and Dworaczyk, B., combinFormation: Mixed-initiative composition of image and text surrogates promotes information discovery. *ACM Transactions on Informations Systems*, 27(1), 1–45, 2008. DOI: 10.1145/1416950.1416955 Cited on page(s) 62

[98] Koenemann, J. and Belkin, N.J., A case for interaction: a study of interactive information retrieval behavior and effectiveness. In *Proceedings of the SIGCHI conference on Human factors in computing systems*, ACM, New York, NY, USA, 205–212, 1996. DOI: 10.1145/238386.238487 Cited on page(s) 25, 26

[99] Koester, H.H. and Levine, S.P., Validation of a keystroke-level model for a text entry system used by people with disabilities. In *Proceedings of the first annual ACM conference on Assistive technologies*, ACM, New York, NY, USA, 115–122, 1994. DOI: 10.1145/191028.191061 Cited on page(s) 96

[100] Kohonen, T., Self-Organizing Maps. Springer, 2001. Cited on page(s) 58

[101] Kotelly, B., Resonance: Introducing the concept of penalty-free deep look ahead with dynamic summarization of arbitrary results sets. In *Workshop on Human-Computer Interaction and Information Retrieval*, MIT CSAIL, Cambridge, Massachusetts, USA, 2007. Cited on page(s) 57

[102] Krichmar, A., Command language ease of use: a comparison of DIALOG and ORBIT. *Online Information Review*, 5(3), 227–240, 1981. DOI: 10.1108/eb024061 Cited on page(s) 17

[103] Kuhlthau, C.C., Inside the search process: Information seeking from the user's perspective. *Journal of the American Society for Information Science*, 42(5), 361–371, 1991. DOI: 10.1002/(SICI)1097-4571(199106)42:5%3C361::AID-ASI6%3E3.0.CO;2-%23 Cited on page(s) 2, 12, 95, 96

[104] Kules, B. and Capra, R., The influence of search stage on gaze behavior in a faceted search interface. In *Proceedings of the 73rd ASIS&T Annual Meeting on Navigating Streams in an Information Ecosystem - Volume 47*, American Society for Information Science, Pittsburgh, Pennsylvania, 1–2, 2010. DOI: 10.1002/meet.14504701398 Cited on page(s) 95

[105] Kules, B., Capra, R., Banta, M. and Sierra, T., What do exploratory searchers look at in a faceted search interface? In *Proceedings of the 9th ACM/IEEE-CS joint conference on Digital libraries*, ACM, Austin, TX, USA, 313–322, 2009. DOI: 10.1145/1555400.1555452 Cited on page(s) 95

[106] Kules, B., Kustanowitz, J. and Shneiderman, B., Categorizing web search results into meaning-ful and stable categories using Fast-Feature techniques. In *Proceedings of the 6th ACM/IEEE-CS joint conference on Digital libraries, ACM/IEEE, Chapel Hill, North Carolina, United States,* 210–219, 2006 DOI: 10.1145/1141753.1141801 Cited on page(s) 38

[107] Kules, B. and Shneiderman, B., Users can change their web search tactics: Design guide-lines for categorized overviews. *Information Processing & Management*, 44(2), 463–484, 2008. DOI: 10.1016/j.ipm.2007.07.014 Cited on page(s) 38

[108] Kunz, C. and Botsch, V., Visual representation and contextualization of search results: list and matrix browser. In *Proceedings of the 2002 international conference on Dublin core and metadata applications: Metadata for e-communities: supporting diversity and convergence*, Dublin Core Metadata Initiative, Florence, Italy, 229–234, 2002. Cited on page(s) 63

[109] Kuo, B.Y.-L., Hentrich, T., Good, B.M. and Wilkinson, M.D., Tag clouds for summarizing web search results. In *Proceedings of the 16th international conference on World Wide Web*, ACM, Banff, Alberta, Canada, 1203–1204, 2007. DOI: 10.1145/1242572.1242766 Cited on page(s) 40

[110] Lalmas, M., XML Retrieval. *Synthesis Lectures on Information Concepts, Retrieval, and Services*, 1(1), 1–111, 2009. DOI: 10.2200/S00203ED1V01Y200907ICR007 Cited on page(s) 91

[111] Lamping, J., Rao, R. and Pirolli, P., A focus+context technique based on hyperbolic geometry for visualizing large hierarchies. In *Proceedings of the SIGCHI conference on Human factors in computing systems*, ACM Press/Addison-Wesley Publishing Co., Denver, Colorado, United States, 401–408, 1995. DOI: 10.1145/223904.223956 Cited on page(s) 60

[112] Lee, S., Buxton, W. and Smith, K.C., A multi-touch three dimensional touch-sensitive tablet. In *Proceedings of the SIGCHI conference on Human factors in computing systems*, ACM, San

Francisco, California, United States, 21–25, 1985. DOI: 10.1145/317456.317461 Cited on page(s) 11

[113] Leuken, R.H.v., Garcia, L., Olivares, X. and Zwol, R.v., Visual diversification of image search results. In *Proceedings of the 18th international conference on World wide web*, ACM, Madrid, Spain, 341–350, 2009. DOI: 10.1145/1526709.1526756 Cited on page(s) 34

[114] Li, J., Huffman, S. and Tokuda, A., Good abandonment in mobile and PC internet search. In *Proceedings of the 32nd international ACM SIGIR conference on Research and development in information retrieval*, ACM, Boston, MA, USA, 43–50, 2009. DOI: 10.1145/1571941.1571951 Cited on page(s) 70, 95

[115] Liaw, S.-S. and Huang, H.-M., An investigation of user attitudes toward search engines as an information retrieval tool. *Computers in Human Behavior*, 19(6), 751–765, 2003. DOI: 10.1016/S0747-5632(03)00009-8 Cited on page(s) 10

[116] Lin, E., Greenberg, S., Trotter, E., Ma, D. and Aycock, J., Does domain highlighting help people identify phishing sites? In *Proceedings of the 2011 annual conference on Human factors in computing systems*, ACM, Vancouver, BC, Canada, 2075–2084, 2011. DOI: 10.1145/1978942.1979244 Cited on page(s) 50

[117] Liu, J. and Belkin, N.J., Personalizing information retrieval for multi-session tasks: the roles of task stage and task type. In *Proceeding of the 33rd international ACM SIGIR conference on Research and development in information retrieval*, ACM, Geneva, Switzerland, 26–33, 2010. DOI: 10.1145/1835449.1835457 Cited on page(s) 76, 95

[118] Mackinlay, J.D., Robertson, G.G. and Card, S.K., The perspective wall: detail and context smoothly integrated. In *Proceedings of the SIGCHI conference on Human factors in computing systems: Reaching through technology*, ACM, New Orleans, Louisiana, United States, 173–176, 1991. DOI: 10.1145/108844.108870 Cited on page(s) 64

[119] Magennis, M. and Rijsbergen, C.J.v., The potential and actual effectiveness of interactive query expansion. In *Proceedings of the 20th annual international ACM SIGIR conference on Research and development in information retrieval*, ACM, Philadelphia, Pennsylvania, United States, 324–332, 1997. DOI: 10.1145/258525.258603 Cited on page(s) 27

[120] Marchionini, G., Exploratory search: from finding to understanding. *Communications of the ACM*, 49(4), 41–46, 2006. DOI: 10.1145/1121949.1121979 Cited on page(s) 2

[121] Marchionini, G., Information Seeking in Electronic Environments. Cambridge University Press, 1995. Cited on page(s) 2, 12, 96

[122] Marshall, C.C., Reading and Writing the Electronic Book. *Synthesis Lectures on Information Concepts, Retrieval, and Services*, 1(1), 1–185, 2009. DOI: 10.2200/S00215ED1V01Y200907ICR009 Cited on page(s) 91

[123] McAlpine, G. and Ingwersen, P., Integrated information retrieval in a knowledge worker support system. In *Proceedings of the 12th annual international ACM SIGIR conference on Research and development in information retrieval*, ACM, Cambridge, Massachusetts, United States, 48–57, 1989. DOI: 10.1145/75334.75341 Cited on page(s) 21, 22

[124] McKenzie, B. and Cockburn, A., An Empirical Analysis of Web Page Revisitation. In *Proceedings of the 34th Annual Hawaii International Conference on System Sciences*, IEEE Computer Society, 5019, 2001. DOI: 10.1109/HICSS.2001.926533 Cited on page(s) 88

[125] Medlock, M., Wixon, D., Terrano, M., Romero, R. and Fulton, B., Using the RITE method to improve products; a definition and a case study. In *Proceedings of Usability Professionals Association*, Orlando, Florida, USA, 2002. Cited on page(s) 11

[126] Modjeska, D.K., Hierarchical data visualization in desktop virtual reality, University of Toronto, 192.2000. Cited on page(s) 67

[127] Morgan, R. and Wilson, M.L., The Revisit Rack: grouping web search thumbnails for optimal visual recognition. In *Proceedings of the 73rd ASIS&T Annual Meeting on Navigating Streams in an Information Ecosystem*, American Society for Information Science, Pittsburgh, Pennsylvania, 1–4, 2010. DOI: 10.1002/meet.14504701176 Cited on page(s) 54

[128] Morris, M.R., A survey of collaborative web search practices. In *Proceeding of the twenty-sixth annual SIGCHI conference on Human factors in computing systems*, ACM, Florence, Italy, 1657–1660, 2008. DOI: 10.1145/1357054.1357312 Cited on page(s) 82

[129] Morris, M.R. and Horvitz, E., SearchTogether: an interface for collaborative web search. In *Proceedings of the 20th annual ACM symposium on User interface software and technology*, ACM, Newport, Rhode Island, USA, 3–12, 2007. DOI: 10.1145/1294211.1294215 Cited on page(s) 82, 83

[130] Morris, M.R. and Teevan, J., Collaborative Web Search: Who, What, Where, When, and Why. *Synthesis Lectures on Information Concepts, Retrieval, and Services*, 1(1), 1–99, 2009. DOI: 10.2200/S00230ED1V01Y200912ICR014 Cited on page(s) 82

[131] Morris, M.R., Teevan, J. and Panovich, K., What do people ask their social networks, and why?: a survey study of status message Q&A behavior. In *Proceedings of the 28th international conference on Human factors in computing systems*, ACM, Atlanta, Georgia, USA, 1739–1748, 2010. DOI: 10.1145/1753326.1753587 Cited on page(s) 78

[132] Morville, P. and Callender, J., Search patterns: Design for Discovery. O'Reilly Media, Inc., 2010. Cited on page(s) 102

[133] Mostafa, J., Personalization in Information Retrieval. Morgan & Claypool Publishers, 2008. Cited on page(s) 74

[134] Munzner, T., H3: laying out large directed graphs in 3D hyperbolic space. In *Proceedings of the 1997 IEEE Symposium on Information Visualization (InfoVis '97)*, IEEE Computer Society, 2, 1997. DOI: 10.1109/INFVIS.1997.636718 Cited on page(s) 66

[135] Nation, D.A., WebTOC: a tool to visualize and quantify Web sites using a hierarchical table of contents browser. In *CHI 98 conference summary on Human factors in computing systems*, ACM, Los Angeles, California, United States, 185–186, 1998. DOI: 10.1145/286498.286664 Cited on page(s) 33

[136] Naveed, N., Grotton, T., Kunegis, J. and Alhadi, A.C., Bad News Travel Fast: A Content-based Analysis of Interestingness on Twitter. In *Proceedings of the ACM WebSci'11*, ACM, Koblenz, Germany, 1–7, 2011. Cited on page(s) 84

[137] Nielsen, J., Heuristic evaluation. in Nielsen, J. and Mack, R.L. eds. *Usability inspection methods*, John Wiley & Sons, Inc., 1994, 25–62. Cited on page(s) 11, 13, 96

[138] Nielsen, J., Usability inspection methods. In *Conference companion on Human factors in computing systems*, ACM, Boston, Massachusetts, United States, 413–414, 1994. DOI: 10.1145/259963.260531 Cited on page(s) 96

[139] Nielsen, J. and Loranger, H., Prioritizing web usability. New Riders Press, 2006. Cited on page(s) 30

[140] Nielsen, J. and Molich, R., Heuristic evaluation of user interfaces. In *Proceedings of the SIGCHI conference on Human factors in computing systems: Empowering people*, ACM, Seattle, Washington, United States, 249–256, 1990. DOI: 10.1145/97243.97281 Cited on page(s) 13, 96

[141] Nowell, L.T., France, R.K. and Hix, D., Exploring search results with Envision. In *CHI '97 extended abstracts on Human factors in computing systems: looking to the future*, ACM, Atlanta, Georgia, 14–15, 1997. DOI: 10.1145/1120212.1120223 Cited on page(s) 63

[142] O'Brien, H.L. and Toms, E.G., The development and evaluation of a survey to measure user engagement. *Journal of the American Society for Information Science and Technology*, 61(1), 50–69, 2010. DOI: 10.1002/asi.v61:1 Cited on page(s) 87, 95

[143] Oddy, R.N., Information retrieval through man-machine dialogue. *Journal of Documentation*, 33(1), 1–14, 1977. DOI: 10.1108/eb026631 Cited on page(s) 17

[144] Oulasvirta, A., Hukkinen, J.P. and Schwartz, B., When more is less: the paradox of choice in search engine use. In *Proceedings of the 32nd international ACM SIGIR conference on Research and development in information retrieval*, ACM, Boston, MA, USA, 516–523, 2009. DOI: 10.1145/1571941.1572030 Cited on page(s) 48

[145] Paas, F., Renkl, A. and Sweller, J., Cognitive Load Theory and Instructional Design: Recent Developments. *Educational Psychologist*, 38(1), 1–4, 2003. DOI: 10.1207/S15326985EP3801_1 Cited on page(s) 21

[146] Paek, T., Dumais, S. and Logan, R., WaveLens: a new view onto Internet search results. In *Proceedings of the 2004 conference on Human factors in computing systems*, ACM, Vienna, Austria, 727–734, 2004. DOI: 10.1145/985692.985784 Cited on page(s) 51

[147] Palay, A.J. and Fox, M.S., Browsing through databases. In *Proceedings of the 3rd annual ACM conference on Research and development in information retrieval*, Butterworth & Co., Cambridge, England, 310–324, 1981. Cited on page(s) 19, 20

[148] Paul, S.A., Hong, L. and Chi, E.H., Is Twitter a Good Place for Asking Questions? A Characterization Study. In *Fifth International AAAI Conference on Weblogs and Social Media*, AAAI, Barcelona, Spain, 2011. Cited on page(s) 78

[149] Paul, S.A. and Morris, M.R., CoSense: enhancing sensemaking for collaborative web search. In *Proceedings of the 27th international conference on Human factors in computing systems*, ACM, Boston, MA, USA, 1771–1780, 2009. DOI: 10.1145/1518701.1518974 Cited on page(s) 82

[150] Pejtersen, A.M., A library system for information retrieval based on a cognitive task analysis and supported by an icon-based interface. In *Proceedings of the 12th annual international ACM SIGIR conference on Research and development in information retrieval*, ACM, Cambridge, Massachusetts, United States, 40–47, 1989. DOI: 10.1145/75334.75340 Cited on page(s) 24, 25

[151] Pirolli, P., Schank, P., Hearst, M. and Diehl, C., Scatter/gather browsing communicates the topic structure of a very large text collection. In *Proceedings of the SIGCHI conference on Human factors in computing systems: common ground*, ACM, Vancouver, British Columbia, Canada, 213–220, 1996. DOI: 10.1145/238386.238489 Cited on page(s) 11, 13, 95

[152] Plaisant, C., Milash, B., Rose, A., Widoff, S. and Shneiderman, B., LifeLines: visualizing personal histories. In *Proceedings of the SIGCHI conference on Human factors in computing systems: common ground*, ACM, Vancouver, British Columbia, Canada, 221–227, 1996. DOI: 10.1145/238386.238493 Cited on page(s) 64

[153] Pratt, W., Dynamic organization of search results using the UMLS. *Proc AMIA Annu Fall Symp.* 480–484, 1997. Cited on page(s) 36

[154] Preece, J., Rogers, Y. and Sharp, H., Beyond Interaction Design: Beyond Human-Computer Interaction. John Wiley & Sons, Inc., New York, NY, USA, 2001. Cited on page(s) 97

[155] Pruitt, J. and Grudin, J., Personas: practice and theory. In *Proceedings of the 2003 conference on Designing for user experiences*, ACM, San Francisco, California, 1–15, 2003. DOI: 10.1145/997078.997089 Cited on page(s) 96

[156] Ranganathan, S.R., Colon Classification. Madras Library Association, 1952. Cited on page(s) 17

[157] Ringel, M., Cutrell, E., Dumais, S. and Horvitz, E., Milestones in time: The value of landmarks in retrieving information from personal stores. In *Proceedings of INTERACT 2003*, Springer-Verlag, 184–191, 2003. Cited on page(s) 89

[158] Robertson, G., Cameron, K., Czerwinski, M. and Robbins, D., Polyarchy visualization: visualizing multiple intersecting hierarchies. In *Proceedings of the SIGCHI conference on Human factors in computing systems: Changing our world, changing ourselves*, ACM, Minneapolis, Minnesota, USA, 423–430, 2002. DOI: 10.1145/503376.503452 Cited on page(s) 73

[159] Robertson, G., Czerwinski, M., Larson, K., Robbins, D.C., Thiel, D. and Dantzich, M.v., Data mountain: using spatial memory for document management. In *Proceedings of the 11th annual ACM symposium on User interface software and technology*, ACM, San Francisco, CA, USA, 53–162, 1998. DOI: 10.1145/288392.288596 Cited on page(s) 65, 69

[160] Robertson, G.G., Mackinlay, J.D. and Card, S.K., Cone Trees: animated 3D visualizations of hierarchical information. In *Proceedings of the SIGCHI conference on Human factors in computing systems: Reaching through technology*, ACM, New Orleans, Louisiana, USA, 189–194, 1991. DOI: 10.1145/108844.108883 Cited on page(s) 65, 70

[161] Robertson, S., Walker, S., Jones, S., Hancock-Beaulieu, M. and Gatford, M., In *Proceedings of the 3rd Text Retrival Conference, NIST,* Gaithersburg, MD, USA, 109–126, 1995 Cited on page(s) 2

[162] Rocchio, J., Relevance feedback in information retrieval. in Salton, G. ed. *The SMART Retrieval System: Experiments in Automatic Document Processing*, Prentice Hall, 1971, 313–323. Cited on page(s) 25

[163] Ruger , S., Multimedia Information Retrieval. *Synthesis Lectures on Information Concepts, Retrieval, and Services*, 1(1), 1–171, 2009. DOI: 10.2200/S00244ED1V01Y200912ICR010 Cited on page(s) 91

[164] Ruthven, I., The context of the interface. In *Proceedings of the 2nd International Symposium on Information Interaction in Context*, London, UK, (Keynote), 2008. DOI: 10.1145/1414694.1414697 Cited on page(s) 86

[165] Salton, G. and Buckley, C., Improving retrieval performance by relevance feedback. *Journal of the American Society for Information Science*, 41(4), 288–297, 1990. DOI: 10.1002/(SICI)1097-4571(199006)41:4%3C288::AID-ASI8%3E3.0.CO;2-H Cited on page(s) 26

[166] Saracevic, T., Relevance reconsidered. In *Proceedings of the 2nd International Conference on the Conceptions of Library and Information Science*, Copenhagen, Denmark, 15, 1996. Cited on page(s) 2, 94

[167] Saracevic, T., The stratified model of information retrieval interaction: extension and applications. In *Proceedings of the Annual Meeting of the American Society for Information Science*, ASIS, Washington DC, USA, 313–327, 1997. Cited on page(s) 9

[168] schraefel, m.c., Karam, M. and Zhao, S., Listen to the Music: Audio Preview Cues for the Exploration of Online Music. In *Proceedings of Interact'03*, Zurich, Switzerland, 192–199, 2003. Cited on page(s) 20, 56

[169] schraefel, m.c., Wilson, M.L., Russell, A. and Smith, D.A., mSpace: improving information access to multimedia domains with multimodal exploratory search. *Communications of the ACM*, 49(4), 47–49, 2006. DOI: 10.1145/1121949.1121980 Cited on page(s) 37

[170] Schrammel, J., Leitner, M. and Tscheligi, M., Semantically structured tag clouds: an empirical evaluation of clustered presentation approaches. In *Proceedings of the 27th international conference on Human factors in computing systems*, ACM, Boston, MA, USA, 2037–2040, 2009. DOI: 10.1145/1518701.1519010 Cited on page(s) 61

[171] Schwarz, J. and Morris, M., Augmenting web pages and search results to support credibility assessment. In *Proceedings of the 2011 annual conference on Human factors in computing systems*, ACM, Vancouver, BC, Canada, 1245–1254, 2011. DOI: 10.1145/1978942.1979127 Cited on page(s) 50

[172] Sebrechts, M.M., Cugini, J.V., Laskowski, S.J., Vasilakis, J. and Miller, M.S., Visualization of search results: a comparative evaluation of text, 2D, and 3D interfaces. In *Proceedings of the 22nd annual international ACM SIGIR conference on Research and development in information retrieval*, ACM, Berkeley, California, United States, 3–10, 1999. DOI: 10.1145/312624.312634 Cited on page(s) 66

[173] Shah, C. and Marchionini, G., Awareness in collaborative information seeking. *Journal of the American Society for Information Science and Technology*, 61(10), 1970–1986, 2010. DOI: 10.1002/asi.21379 Cited on page(s) 82

[174] Shneiderman, B., Tree visualization with tree-maps: 2-d space-filling approach. *ACM Transactions on Graphics*, 11(1), 92–99, 1992. DOI: 10.1145/102377.115768 Cited on page(s) 63, 66

[175] Shneiderman, B., Feldman, D., Rose, A. and Grau, X.F., Visualizing digital library search results with categorical and hierarchial axes. In *Proceedings of the Fifth ACM International Conference on Digital Libraries*, ACM, San Antonio, TX, USA, 57–66, 2000. DOI: 10.1145/336597.336637 Cited on page(s) 63, 67

[176] Shneiderman, B., The future of interactive systems and the emergence of direct manipulation. *Behaviour & Information Technology*, 1(3), 237–256, 1982. Cited on page(s) 11

[177] Shute, S.J. and Smith, P.J., Knowledge-Based Search Tactics. *Information Processing & Management*, 29(1), 29–45, 1993. DOI: 10.1016/0306-4573(93)90021-5 Cited on page(s) 12, 96

[178] Singer, E., Van Hoewyk, J. and Maher, M.P., Does the payment of incentives create expectation effects? *Public Opinion Quarterly*, 62(2), 152, 1998. DOI: 10.1086/297838 Cited on page(s) 94

[179] Slonim, J., Maryanski, F.J. and Fisher, P.S., Mediator: An integrated approach to Information Retrieval. In *Proceedings of the 1st annual international ACM SIGIR conference on Information storage and retrieval*, ACM, 14–36, 1978. DOI: 10.1145/800096.803134 Cited on page(s) 17, 18

[180] Sohn, T., Li, K.A., Griswold, W.G. and Hollan, J.D., A diary study of mobile information needs. In *Proceeding of the twenty-sixth annual SIGCHI conference on Human factors in computing systems*, ACM, Florence, Italy, 433–442, 2008. DOI: 10.1145/1357054.1357125 Cited on page(s) 87

[181] Spoerri, A., Infocrystal: a visual tool for information retrieval. In *Proceedings of the 4th conference on Visualization '93*, IEEE Computer Society, San Jose, California, 150–157, 1993. DOI: 10.1109/VISUAL.1993.398863 Cited on page(s) 58, 62

[182] Stein, A. and Thiel, U., A Conversational Model of Multimodal Interaction in Information Systems. In *Proceedings of the 11 National Conference on Artificial Intelligence*, AAAI Press/MIT Press, Washington DC, 283–288, 1993. Cited on page(s) 17

[183] Stoica, E. and Hearst, M.A., Nearly-automated metadata hierarchy creation. In *Proceedings of HLT-NAACL 2004: Short Papers*, Association for Computational Linguistics, Boston, Massachusetts, 117–120, 2004. DOI: 10.3115/1613984.1614014 Cited on page(s) 35

[184] Su, L.T., Evaluation measures for interactive information retrieval. *Information Processing & Management*, 28(4), 503–516, 1992. DOI: 10.1016/0306-4573(92)90007-M Cited on page(s) 94

[185] Sushmita, S., Lalmas, M. and Tombros, A., Using digest pages to increase user result space: Preliminary designs. In *SIGIR Workshop on Aggregated Search*, 2008. Cited on page(s) 49

[186] Tauscher, L. and Greenberg, S., How people revisit web pages: empirical findings and implications for the design of history systems. *International Journal of Human-Computer Studies*, 47(1), 97–137, 1997. DOI: 10.1006/ijhc.1997.0125 Cited on page(s) 88

[187] Taylor, R.S., Question-negotiation and Information Seeking in Libraries. *College & Research Libraries*, 29(178–194, 1968. Cited on page(s) 17

[188] Teevan, J., The re:search engine: simultaneous support for finding and re-finding. In *Proceedings of the 20th annual ACM symposium on User interface software and technology*, ACM, Newport, Rhode Island, USA, 23–32, 2007. DOI: 10.1145/1294211.1294217 Cited on page(s) 89

[189] Teevan, J., Alvarado, C., Ackerman, M.S. and Karger, D.R., The perfect search engine is not enough: a study of orienteering behavior in directed search. In *Proceedings of the SIGCHI conference on Human factors in computing systems*, ACM, Vienna, Austria, 415–422, 2004. DOI: 10.1145/985692.985745 Cited on page(s) 89

[190] Teevan, J., Cutrell, E., Fisher, D., Drucker, S.M., Ramos, G., André, P. and Hu, C., Visual snippets: summarizing web pages for search and revisitation. In *Proceedings of the 27th international conference on Human factors in computing systems*, ACM, Boston, MA, USA, 2023–2032, 2009. DOI: 10.1145/1518701.1519008 Cited on page(s) 53, 54, 55

[191] Teevan, J., Dumais, S.T., Liebling, D.J. and Hughes, R.L., Changing how people view changes on the web. In *Proceedings of the 22nd annual ACM symposium on User interface software and technology*, ACM, Victoria, BC, Canada, 237–246, 2009. DOI: 10.1145/1622176.1622221 Cited on page(s) 89

[192] Teevan, J., Ramage, D. and Morris, M.R., #TwitterSearch: a comparison of microblog search and web search. In *Proceedings of the fourth ACM international conference on Web search and data mining*, ACM, Hong Kong, China, 35–44, 2011. DOI: 10.1145/1935826.1935842 Cited on page(s) 83

[193] Teskey, F.N., Intelligent support for interface systems. In *Proceedings of the 11th annual international ACM SIGIR conference on Research and development in information retrieval*, ACM, Grenoble, France, 401–415, 1988. Cited on page(s) 23

[194] Tombros, A., Villa, R. and Van Rijsbergen, C.J., The effectiveness of query-specific hierarchic clustering in information retrieval. *Information Processing & Management*, 38(4), 559–582, 2002. DOI: 10.1016/S0306-4573(01)00048-6 Cited on page(s) 13

[195] Tufte, E.R., Envisioning information. Graphics Press Cheshire, 1990. Cited on page(s) 63

[196] Tunkelang, D., Faceted Search. *Synthesis Lectures on Information Concepts, Retrieval, and Services*, 1(1), 1–80, 2009. DOI: 10.2200/S00190ED1V01Y200904ICR005 Cited on page(s) 40

[197] Turetken, O. and Sharda, R., Clustering-Based Visual Interfaces for Presentation of Web Search Results: An Empirical Investigation. *Information Systems Frontiers*, 7(3), 273–297, 2005. DOI: 10.1007/s10796-005-2770-7 Cited on page(s) 35

[198] Veerasamy, A. and Belkin, N.J., Evaluation of a tool for visualization of information retrieval results. In *Proceedings of the 19th annual international ACM SIGIR conference on Research and development in information retrieval*, ACM, Zurich, Switzerland, 85–92, 1996. DOI: 10.1145/243199.243218 Cited on page(s) 58, 61

[199] Vermeeren, A.P.O.S., Law, E.L.-C., Roto, V., Obrist, M., Hoonhout, J. and Väänänen-Vainio-Mattila, K., User experience evaluation methods: current state and development needs. In *Proceedings of the 6th Nordic Conference on Human-Computer Interaction: Extending Boundaries*, ACM, Reykjavik, Iceland, 521–530, 2010. DOI: 10.1145/1868914.1868973 Cited on page(s) 11

[200] Vredenburg, K., Mao, J.-Y., Smith, P.W. and Carey, T., A survey of user-centered design practice. In *Proceedings of the SIGCHI conference on Human factors in computing systems: Changing our world, changing ourselves*, ACM, Minneapolis, Minnesota, USA, 471–478, 2002. DOI: 10.1145/503376.503460 Cited on page(s) 11

[201] Wharton, C., Rieman, J., Lewis, C. and Polson, P., The cognitive walkthrough method: A practitioner's guide. in Nielsen, J. and Mack, R.L. eds. *Usability inspection methods*, John Wiley & Sons, Inc., 1994, 105–140. Cited on page(s) 96

[202] White, R. and Roth, R., Exploratory Search: Beyond the Query-Response Paradigm. Morgan & Claypool, 2009. DOI: 10.2200/S00174ED1V01Y200901ICR003 Cited on page(s) 2, 84, 87

[203] White, R.W. and Drucker, S.M., Investigating behavioral variability in web search. In *Proceedings of the 16th international conference on World Wide Web*, ACM, Banff, Alberta, Canada, 21–30, 2007. DOI: 10.1145/1242572.1242576 Cited on page(s) 86

[204] White, R.W., Jose, J.M. and Ruthven, I., A task-oriented study on the influencing effects of query-biased summarisation in web searching. *Information Processing & Management*, 39(5), 707–733, 2003. DOI: 10.1016/S0306-4573(02)00033-X Cited on page(s) 51

[205] Wilson, M., Russell, A., schraefel, m.c. and Smith, D.A., mSpace mobile: a UI gestalt to support on-the-go info-interaction. In *CHI '06 extended abstracts on Human factors in computing systems*, ACM, Montréal, Québec, Canada, 247–250, 2006. Cited on page(s) 87

[206] Wilson, M.J. and Wilson, M.L., Tag clouds and keyword clouds: evaluating zero-interaction benefits. In *Proceedings of the 2011 annual conference extended abstracts on Human factors in computing systems*, ACM, Vancouver, BC, Canada, 2383–2388, 2011. DOI: 10.1145/1979742.1979913 Cited on page(s) 40

[207] Wilson, M.L., An Analytical Inspection Framework for Evaluating the Search Tactics and User Profiles Supported by Information Seeking Interfaces, University of Southampton, 249.2009. Cited on page(s) 12, 96

[208] Wilson, M.L., André, P. and schraefel, m.c., Backward highlighting: enhancing faceted search. In *Proceedings of the 21st annual ACM symposium on User interface software and technology*, ACM, Monterey, CA, USA, 235–238, 2008. DOI: 10.1145/1449715.1449754 Cited on page(s) 11, 38, 94, 95

[209] Wilson, M.L., Kules, B., schraefel, m.c. and Shneiderman, B., From Keyword Search to Exploration: Designing Future Search Interfaces for the Web. *Foundations and Trends in Web Science*, 2(1), 1–97, 2010. DOI: 10.1561/1800000003 Cited on page(s) 102

[210] Wilson, M.L. and schraefel, m.c., A longitudinal study of exploratory and keyword search. In *Proceedings of the 8th ACM/IEEE-CS joint conference on Digital libraries*, ACM, Pittsburgh, PA, USA, 52–56, 2008. DOI: 10.1145/1378889.1378899 Cited on page(s) 95

[211] Wilson, M.L., schraefel, m.c. and White, R.W., Evaluating Advanced Search Interfaces using Established Information-Seeking Models. *Journal of the American Society for Information Science and Technology*, 60(7), 1407–1422, 2009. DOI: 10.1002/asi.21080 Cited on page(s) 12, 96, 97

[212] Winograd, T. and Flores, F., Understanding computers and cognition. Ablex Publishing Corp. Norwood, NJ, USA, 1986. Cited on page(s) 17

[213] Wongsuphasawat, K., Gómez, J.A.G., Plaisant, C., Wang, T.D., Taieb-Maimon, M. and Shneiderman, B., LifeFlow: visualizing an overview of event sequences. In *Proceedings of the 2011 annual conference on Human factors in computing systems*, ACM, Vancouver, BC, Canada, 1747–1756, 2011. DOI: 10.1145/1978942.1979196 Cited on page(s) 65

[214] Woodruff, A., Faulring, A., Rosenholtz, R., Morrsion, J. and Pirolli, P., Using thumbnails to search the Web. In *Proceedings of the SIGCHI conference on Human factors in computing systems*, ACM, Seattle, Washington, United States, 198–205, 2001. DOI: 10.1145/365024.365098 Cited on page(s) 54

[215] Woodruff, A., Rosenholtz, R., Morrison, J.B., Faulring, A. and Pirolli, P., A comparison of the use of text summaries, plain thumbnails, and enhanced thumbnails for Web search tasks. *Journal of the American Society for Information Science and Technology*, 53(2), 172–185, 2002. DOI: 10.1002/asi.10029 Cited on page(s) 54, 55, 56

[216] Wu, H., Kazai, G. and Taylor, M., Book Search Experiments: Investigating IR Methods for the Indexing and Retrieval of Books. in Macdonald, C., Ounis, I., Plachouras, V., Ruthven, I. and White, R. eds. *Advances in Information Retrieval*, Springer Berlin / Heidelberg, 2008, 234–245. DOI: 10.1007/978-3-540-78646-7_23 Cited on page(s) 91

[217] Yee, K.-P., Swearingen, K., Li, K. and Hearst, M., Faceted metadata for image search and browsing. In *Proceedings of the SIGCHI conference on Human factors in computing systems*,

ACM, Ft. Lauderdale, Florida, USA, 401–408, 2003 DOI: 10.1145/642611.642681 Cited on page(s) 11, 36, 48

[218] Yeh, T., White, B., Pedro, J.S., Katz, B. and Davis, L.S., A case for query by image and text content: searching computer help using screenshots and keywords. In *Proceedings of the 20th international conference on World wide web*, ACM, Hyderabad, India, 775–784, 2011. DOI: 10.1145/1963405.1963513 Cited on page(s) 32

[219] Zhang, J. and Marchionini, G., Evaluation and evolution of a browse and search interface: relation browser. In *Proceedings of the 2005 national conference on Digital government*, Digital Government Society of North America, Atlanta, Georgia, 179–188, 2005. Cited on page(s) 38, 68

[220] Zhao, Y. and Karypis, G., Evaluation of hierarchical clustering algorithms for document datasets. In *Proceedings of the eleventh international conference on Information and knowledge management*, ACM, McLean, Virginia, USA, 515–524, 2002. DOI: 10.1145/584792.584877 Cited on page(s) 34

[221] Zheng, X.S., Chakraborty, I., Lin, J.J.-W. and Rauschenberger, R., Correlating low-level image statistics with users - rapid aesthetic and affective judgments of web pages. In *Proceedings of the 27th international conference on Human factors in computing systems*, ACM, Boston, MA, USA, 1–10, 2009. DOI: 10.1145/1518701.1518703 Cited on page(s) 11, 53

Author's Biography

MAX L. WILSON

Max L. Wilson is a Lecturer in HCI and Information Seeking, in the Future Interaction Technology Lab at Swansea University, UK. His research focuses on Search User Interface design, taking a multidisciplinary perspective from both Human-Computer Interaction (the presentation and interaction) and Information Science (the information and seeking behaviours). His doctoral work, which won the best JASIST article in 2009, focused on evaluating Search User Interfaces using models of Information Seeking. Max received his PhD from Southampton University, under the supervision of m.c. schraefel and Dame Wendy Hall, where much of his work was grounded in supporting Exploratory Search, and within the developing context of Web Science.

Max mainly publishes in HCI and Information Science communities, including a monograph with co-authors schraefel, Kules, and Shneiderman on future Search User Interfaces for the Web, and book chapters on Search User Interface design and Casual-Leisure Information Behaviour. Given the social evolution of the Web, and his interest in Web Science, some of Max's more recent work has focused on how people use social media, especially microblogging, as an information source.

Max's research has been published at ACM CHI, ACM UIST, AAAI Conference on Weblogs and Social media, ACM JCDL, and Hypertext, and in journals including JASIST and Information Processing & Management. Max reviews for several journals and conferences, and has been on the organising committee for CHI2010, CHI2011, IliX2010, IliX2012, and ICWSM2012. Max is also leading the RepliCHI project, focusing on how the CHI community currently manages the replicability of HCI research. Finally, Max co-chairs the euroHCIR workshops aimed at stimulating the European communities interested in Human-Computer Interaction and Information Retrieval.

CPSIA information can be obtained at www.ICGtesting.com
Printed in the USA
BVOW060009091211

277913BV00005B/6/P